# THE PERIODIC TABLE
# TABLE
# IN MINUTES

## DAN GREEN

Quercus

# The periodic table

| | | | | | | | | | | | | | | | | | |
|---|---|---|---|---|---|---|---|---|---|---|---|---|---|---|---|---|---|
| **H** 1 Hydrogen | | | | | | | | | | | | | | | | | |
| **Li** 3 Lithium | **Be** 4 Beryllium | | | | | | | | | | | | | | | | |
| **Na** 11 Sodium | **Mg** 12 Magnesium | | | | | | | | | | | | | | | | |
| **K** 19 Potassium | **Ca** 20 Calcium | **Sc** 21 Scandium | **Ti** 22 Titanium | **V** 23 Vanadium | **Cr** 24 Chromium | **Mn** 25 Manganese | **Fe** 26 Iron | **Co** 27 Cobalt | | | | | | | | | |
| **Rb** 37 Rubidium | **Sr** 38 Strontium | **Y** 39 Yttrium | **Zr** 40 Zirconium | **Nb** 41 Niobium | **Mo** 42 Molybdenum | **Tc** 43 Technetium | **Ru** 44 Ruthenium | **Rh** 45 Rhodium | | | | | | | | | |
| **Cs** 55 Caesium | **Ba** 56 Barium | * | **Hf** 72 Hafnium | **Ta** 73 Tantalum | **W** 74 Tungsten | **Re** 75 Rhenium | **Os** 76 Osmium | **Ir** 77 Iridium | | | | | | | | | |
| **Fr** 87 Francium | **Ra** 88 Radium | ** | **Rf** 104 Rutherfordium | **Db** 105 Dubnium | **Sg** 106 Seaborgium | **Bh** 107 Bohrium | **Hs** 108 Hassium | **Mt** 109 Meitnerium | | | | | | | | | |

| | | | | | | | |
|---|---|---|---|---|---|---|---|
| * | **La** 57 Lanthanum | **Ce** 58 Cerium | **Pr** 59 Praseodymium | **Nd** 60 Neodymium | **Pm** 61 Promethium | **Sm** 62 Samarium | **Eu** 63 Europium |
| ** | **Ac** 89 Actinium | **Th** 90 Thorium | **Pa** 91 Protactinium | **U** 92 Uranium | **Np** 93 Neptunium | **Pu** 94 Plutonium | **Am** 95 Americium |

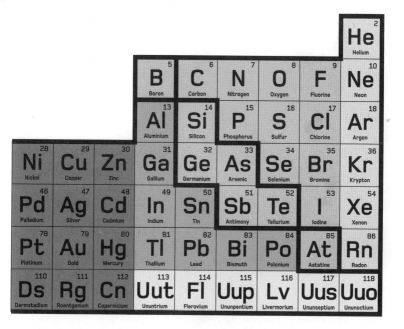

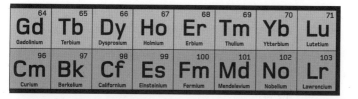

# CONTENTS

# Key to element data

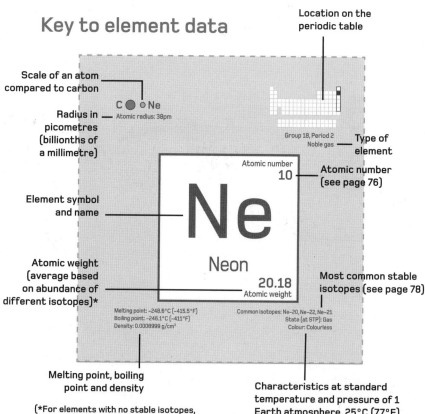

**Location on the periodic table**

**Scale of an atom compared to carbon**

C ● ○ Ne
Atomic radius: 38pm

**Radius in picometres (billionths of a millimetre)**

Group 18, Period 2
Noble gas

**Type of element**

Atomic number
10

**Atomic number (see page 76)**

**Element symbol and name**

Ne

Neon

20.18
Atomic weight

**Atomic weight (average based on abundance of different isotopes)\***

**Most common stable isotopes (see page 78)**

Melting point: -248.6°C (-415.5°F)
Boiling point: -246.1°C (-411°F)
Density: 0.0008999 g/cm³

Common isotopes: Ne-20, Ne-22, Ne-21
State (at STP): Gas
Colour: Colourless

**Melting point, boiling point and density**

**Characteristics at standard temperature and pressure of 1 Earth atmosphere, 25°C (77°F)**

(\*For elements with no stable isotopes, atomic weight of most stable form is listed)

# Introduction

The periodic table is one of the crown jewels of science. The classification of elements is one of the greatest and most highly prized discoveries, and is ranked in the first order of generalizations about our universe. The table's castle-like shape has become an instantly recognizable design icon — a permanent (though not unchanging) feature on the wall of the chemistry lab, and on everything from mugs to novelty ties.

Russian chemist Dmitri Mendeleev invented the periodic table in 1869, using the chemical properties of the elements as an organizing principle. Its principal strength is that it was able to include and explain later findings, such as the internal make up of atoms, the discovery of atomic number and valence theories of chemical bonding. It even predicted the positions of several undiscovered elements. The turreted walls of the table encompass some of science's most important discoveries, including atomic theory, electricity, the spectrum of light and other electromagnetic radiation, radioactivity and quantum physics.

The elements were once defined as the simplest substances of all – the very word comes from the Latin word *elementum*, meaning 'first principle' or 'most basic form'. Today, we know that in fact every atom is composed of even smaller and more basic elementary particles, so a modern definition of an element might be a substance composed of atoms with the same number of subatomic proton particles in their nuclei. The investigation of the nature of matter and atomic structure has been a driving force in the history of science, leading not only to huge and sometimes surprising advances in our understanding of elements and atoms, but also to a wide range of technological applications that have shaped our modern society.

Like the law of gravitation or the theory of evolution, the periodic law, describing how chemical elements are ordered, is held to be universally true, and indeed, the properties of key biological elements formed part of the famous 1974 Arecibo message, humanity's first deliberate message to the stars. If we were ever to encounter an alien culture, this is the sort of common ground we could discuss: although our names for the elements would be different, we could all agree on the 118 different types of atom from which all matter is made.

# Elements and atoms

The Nobel-prize-winning physicist Richard Feynman was once asked what single scientific fact should be preserved in the event of an apocalypse. His response was definitive: that all things are made of atoms – little particles that move around in perpetual motion, attracting each other when they are a little distance apart, but repelling upon being squeezed into one another. The atomic hypothesis is an ancient concept, dating back to the Ancient Greeks. However, evidence for small packages of matter comes from much later experiments in fields such as Brownian motion and atomic force microscopy (see page 116).

Elements are substances that cannot be broken down into simpler components, either by physical or chemical means. Each comprises only one variety of atom, determined by the number of protons in its nucleus (see page 10). Today, there are 118 such types of atom (and many more 'isotopes', see page 78) , 92 of which occur naturally on Earth.

A pure element contains atoms of a single type – for example the copper in electrical wire, the carbon in coal and the helium gas in party balloons. In practice, however, very few everyday materials are pure in this way.

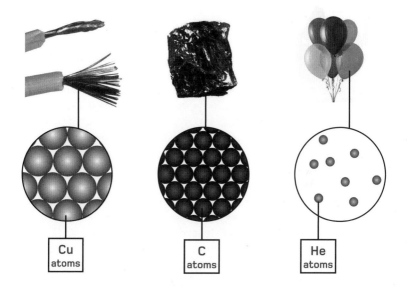

Cu
atoms

C
atoms

He
atoms

# Atomic structure

**M**ost people might have an idea of an atom as a mini solar system, the nucleus at the centre where the Sun would be, orbited by electron 'planets'. The simplicity of this 'planetary model', sketched out by New Zealand physicist Ernest Rutherford in 1911 (see page 72), gives it staying power. Niels Bohr's 1913 quantum description, in which 'probalistic' electrons reside in defined orbitals, is more complete (see page 74).

The atomic nucleus consists of positively charged protons and electrically neutral neutrons, collectively called nucleons. Nearly all of the atom's mass is concentrated here. The number of protons determines the element – even when they lose electrons, atoms still retain their identity. The charge of the protons is neutralized by an equal number of negatively charged electrons, bound by electrostatic attraction to the nucleus and resulting in an overall neutral atom. Atoms are typically about 1 Ångstrom (0.1 nanometres) in diameter, about 100,000 times smaller than a red blood cell.

# Components of a helium atom

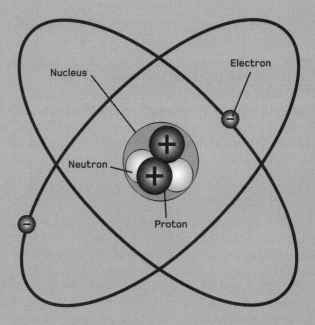

# Subatomic particles

The question of what is 'fundamental' is a core pursuit of science, but the basic unit of matter differs depending on who is asked. For chemists, it is the atom, since this is how matter interacts in everyday conditions. For physicists, however, it's the exotic elementary particles that spray out of high-energy proton collisions that lie at the heart of matter.

Atoms have three basic building blocks – protons, neutrons and electrons. Of these, only the electron is thought to be fundamental. The electron was the first subatomic particle to be found, in 1897 (see page 70), and this was followed by discoveries of the proton in 1919 and the neutron in 1932. The electron belongs to the family of leptons, which also includes muon, tau and neutrino particles. Protons and neutrons are baryons – composite particles made of different combinations of three quarks. These nucleons are heavy and almost all the intrinsic mass of an atom resides in the nucleus with them. Electrons are nearly 2,000 times lighter.

Tracks from subatomic particles
in the Large Hadron Collider

# Electron configuration

Electrons in orbit around an atomic nucleus cannot take up any position they desire. Instead, they occupy atomic orbitals, arranged in 'shells' that are associated with a fixed (quantized) energy. Each shell holds a specific number of electrons and when it is full, electrons start to fill up the next level. According to quantum physics, orbitals are not sharply defined structures, but are instead fuzzy-edged 'zones of probability', in which there is a 95 per cent probability of finding an electron. Gaps between orbitals are where electrons are never to be found.

An element's electron configuration is key to its physical properties and chemical behaviour. The number of electrons in the outermost shell (known as the valence shell) and their energies largely determine the type of bonding and range of compounds and element forms. Electrons in orbitals close to the nucleus are the most tightly bound, and a complete valence shell is a more stable electron configuration. This configuration also explains many of the trends observed in the periodic table (see page 118).

Electrons do not orbit the atomic nucleus like planets orbiting a sun. Instead, they smear out, forming an 'electron cloud' around the nucleus. They are confined in limited spaces, called orbitals, arranged around the nucleus. The shapes of these orbitals are defined by mathematical functions.

# Forces in the nucleus

Hydrogen, the simplest element, has just a single positively charged proton for its nucleus. Heavier elements, however, have many more protons, so how does the nucleus overcome the electrostatic repulsion between these positive electrical charges? The answer lies in the strong force (or 'strong nuclear interaction'), a short-range attractive force that operates over distances of a femtometre ($1 \times 10^{-15}$ m). It not only attracts protons and neutrons to each other, but also binds quarks together to form the nucleons themselves. Another force at work in the nucleus – the weak force – allows neutrons to occasionally turn into protons in radioactive beta decay (see page 18).

The combined mass of nucleons in a nucleus is fractionally lower than their separated masses, and when elements are forged from individual nucleons, this excess mass is liberated as 'binding energy'. Nuclear reactions can tap into this energy source either through fusion (joining light nuclei to make heavier ones), or fission (splitting heavy elements to make lighter ones).

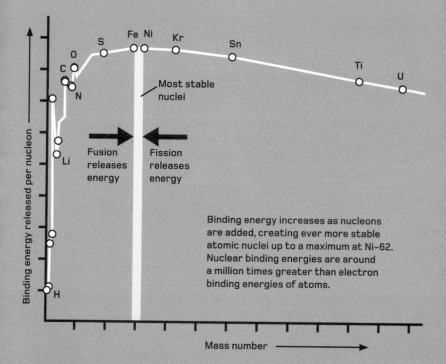

## Nuclear binding energies

Most stable nuclei

Fusion releases energy

Fission releases energy

Binding energy released per nucleon

Mass number

Binding energy increases as nucleons are added, creating ever more stable atomic nuclei up to a maximum at Ni-62. Nuclear binding energies are around a million times greater than electron binding energies of atoms.

# Unstable nuclei and radioactivity

The natural phenomenon of radioactivity came to light in 1896, when Henri Becquerel (1852–1908) noticed that photographic plates placed in a drawer with uranium salts appeared to have exposed. Investigating these mysterious radiations further, he found that some were charged particles. The discovery shook science to the core: if bits could be knocked off the atom, then it could not be the fundamental unit of matter.

The more nucleons a nucleus has, the more unstable it becomes: all elements heavier than bismuth are radioactive. Radiation is matter and/or energy fired out of the nucleus when it decays to become more stable, normally losing mass, and changing to a different element in the process. Ernest Rutherford classified nuclear radiation by penetrating power in three types, from easily stopped heavy alpha particles, with a charge of +2, through fast negatively charged beta particles (electrons released when a proton changes into a neutron) to highly penetrating gamma rays of pure electromagnetic energy.

## Three types of decay

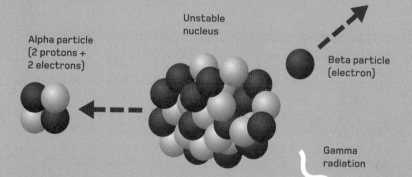

Unstable nucleus

Alpha particle
(2 protons +
2 electrons)

Beta particle
(electron)

Gamma radiation

Radioactive decay occurs when unstable nuclei lose energy. Alpha and beta radiation transform an atom into a different chemical element.

# Molecules and compounds

In the mild conditions that prevail on Earth, most things are not made of solitary atoms, but of combinations of atoms called molecules. Just like the difference between atoms and elements, the distinction between molecules and compounds is subtle but important. Molecules are formed by two or more atoms bound together; compounds are molecules composed of at least two different elements. All compounds are molecules, but not all molecules are compounds. So-called ionic compounds, meanwhile, consist of large numbers of differing atoms bonded together in a continuous mass.

'Molecular elements' consist solely of one element. Several familiar gases exist as diatomic molecules – including oxygen ($O_2$), nitrogen ($N_2$) and hydrogen ($H_2$). Water ($H_2O$), by contrast, is a compound of more than one element (in this case hydrogen and oxygen). Molecules are generally electrically neutral and covalently bonded (see page 28) but they can also lose electrons to become molecular ions (see page 26).

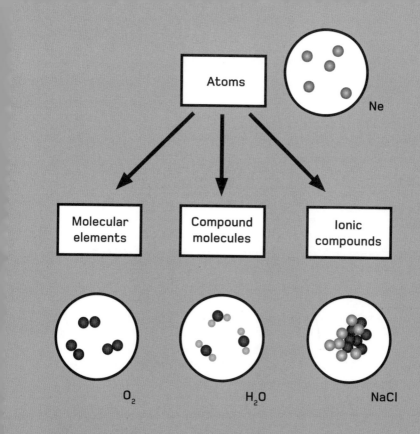

# Solids, liquids and gases

States of matter refer to the form that bulk material takes under normal conditions – typically gaseous, liquid or solid. What counts is whether the consituent particles (atoms, ions or molecules) are linked together or separated.

Solid objects fill space with a fixed 3-D shape. Bonds prevent particles from moving freely, and so solids only change shape when force is applied. Particles in liquids are more mobile, with bonds constantly breaking and reforming so they can flow past each other and take the shape of a container. At any given temperature and pressure, however, their volume is constant, making liquids almost incompressible. Gases also flow like fluids, but with little to no interaction between atoms or molecules, they will expand to fill a container. They can be compressed, and so have no fixed shape or volume. Each state is stable within a range of temperature and pressure regimes. Heating increases the kinetic energy of particles until the point they have enough energy to break free of intermolecular ionic bonds and a change of state occurs.

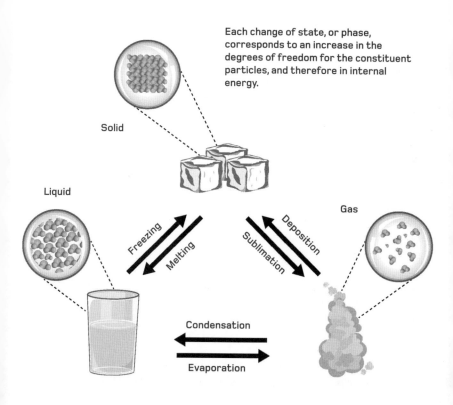

Each change of state, or phase, corresponds to an increase in the degrees of freedom for the constituent particles, and therefore in internal energy.

Solid

Liquid

Gas

Freezing

Melting

Deposition

Sublimation

Condensation

Evaporation

# Ions

In normal matter, positive and negative charge are precisely balanced, so that objects are not usually attracted to one another by electrostatic forces. Ions, however, are atoms or molecules in which the number of positively charged protons and negatively charged electrons are not equal. This charge imbalance provides the opportunity for different interactions of matter through electrostatic attraction and repulsion. 'Anions' carry an excess of electrons and an overall negative charge, while 'cations' have an electron shortfall and a resultant positive charge. The charge on an atom or molecule in ionic form is usually indicated in superscript form: for instance $Cl^-$ and $Mg^{2+}$.

Ions are part of everyday matter, and easy to create. Light knocks electrons off atoms all the time, while dissolved ionic salts release ions into solution. Even walking around on rubber soles can create a build-up of charge, which grounds with a shock when you touch a metal handrail or shake someone's hand. The energy required to remove an electron is called ionization energy (see page 162).

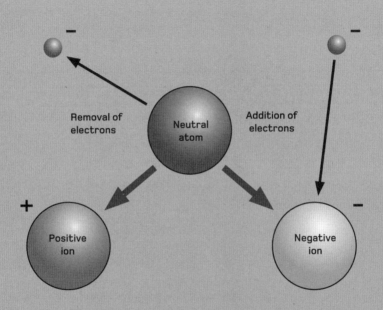

# Ionic bonding

Ionic bonding is a type of chemical bonding in which ions with opposing charges form crystalline solids. It is also known as electrovalent bonding because atoms will often 'exchange' electrons to complete their valence shells (see page 14) and achieve a stable configuration. The resulting ions form compounds held together by electrostatic forces of attraction.

Ionic compounds typically form between metal cations and non-metal anions. A classic example is sodium which bonds with chlorine to form sodium chloride ($NaCl$). The more electronegative (see page 160) an element, the more ionic the nature of the bond. However, there is often some covalent (electron-sharing) character to ionic compounds. As well as individual ions, covalently bonded molecules, such as carbonate ($CO_3^{2-}$) can form polyatomic ions. Ions pack into the smallest possible space, building up into a crystal lattice of regularly repeating units. Ionic compounds generally have high melting points and in their solid form are usually brittle and tend not to conduct electricity.

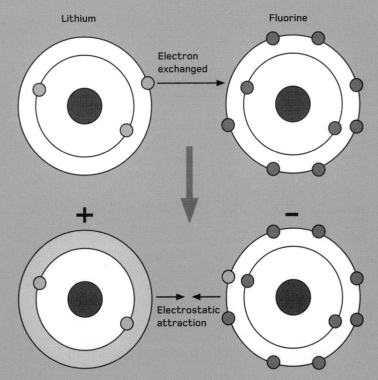

Ionic bonding in lithium fluoride (LiF)

Lithium

Fluorine

Electron
exchanged

+

−

Electrostatic
attraction

# Covalent bonding

**C**ovalent bonding is a model of chemical bonding where electrons are 'shared' rather than exchanged. Atoms will 'seek' to fill their outermost shell to acheive a more stable electron configuration. For example four hydrogen atoms, each with one electron and 'space' for one more in the outer shell, can bond with one carbon atom – carrying four outer electrons, but with room for eight in its valence shell – to form methane ($CH_4$).

'Going Dutch' on electrons allows a variety of bonding options. Many atoms share more than one electron to form double and triple bonds. Covalent compounds form mostly between non-metals, and since there is little opportunity for charge to flow when electrons are tightly bound, they are typically electrical insulators. Neutral molecules experience little mutual attraction, so many such compounds are gases and low-boiling-point liquids, such as carbon dioxide and pentane. Larger 'macromolecules' bond covalently in three dimensions, forming hard, high-melting-point solids, such as diamond.

# Covalent bonding in fluoride gas ($F_2$)

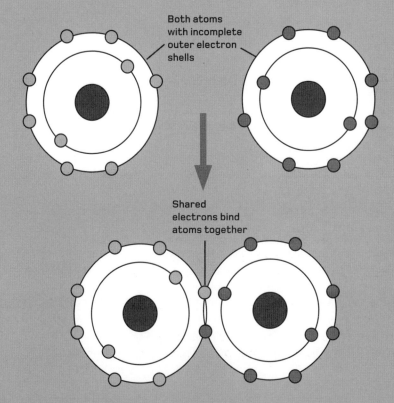

Both atoms with incomplete outer electron shells

Shared electrons bind atoms together

# Metallic bonding

A third class of bonding in compounds is metallic bonding. It is often claimed that this type of bonding is typical of metals. Rather, it is the properties of the metal elements (see page 164) that arise from their tendency to form these bonds. Bonding in metals has characteristics of both ionic and covalent bonding. Rather than being associated with any particular atom, the valence electrons are free to move around multiple positively charged 'atomic cores'.

The electrostatic attraction holding the solid together recalls the ionic bond, while the 'delocalized' electrons are 'shared' between atoms in overlapping orbitals, as in covalent bonds. A positively charged core is not an ion, but is made up of the atomic nucleus and all the electrons that aren't in the outer shell. This type of bonding is very strong and takes a great deal of energy to break – this is why metals, in general, have high melting and boiling points. It also allows metals great flexibility, unlike the brittle rigidity of ionic or covalent compounds.

# Metallic bonding in sodium

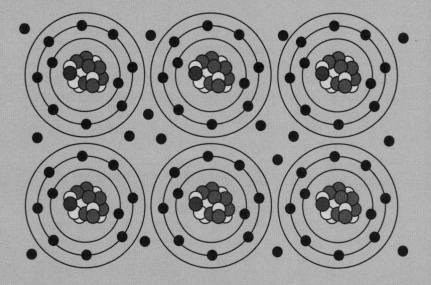

Sharing electrons across positively charged ions results in a metallic crystal lattice, in which the electrons are free to move and conduct electricity.

# Chemical reactions

The school textbook account of a chemical reaction involves the change of a substance from one thing to another. This wide-ranging definition takes in the huge variety of transitions that happen in nature, but does not provide any clues as to how or why things react, and what determines the products.

Chemical reactions conventionally refer to interactions involving valence electrons – breaking bonds between substances and forming new ones. This covers a panopoly of changes from simple combinations, to replacement of individual atoms or groups. Breaking chemical bonds requires energy, and a reaction will proceed spontaneously if the new bonds made release a *greater* amount of energy when they form. Otherwise, chemical change needs to be encouraged with the provision of energy. Catalysts, however, lower the energy required for a reaction or allow it to proceed more quickly. Considering reaction energetics allows reactions to be predicted and manipulated, and makes multi-step syntheses possible.

Fireworks employ a complex series of simple chemical reactions: colours arise from combustion reactions of metal salts.

# Structural formulae

A chemical formula is a partial picture of a molecule. Although it is written in words, it conveys an idea of the constituent elements in a compound and how they might be connected together. Thus, $CH_3COOH$ – also known as acetic acid, or white vinegar – is a methyl group (one carbon atom bonded to three hydrogen atoms, $CH_3$), connected to a carboxylic acid group (a carbon atom double bonded to an oxygen atom and a hydroxide OH group). However, there are many things that this type of formula alone can't tell us – how the molecule is distributed in space, for example. Structural formulae, on the other hand, really *are* drawings, able to show the hexagonal double bonds of a benzene ring ($C_6H_6$, opposite), if only in two dimensions. John Dalton (see page 60) made the first attempt at a standardized chemical nomenclature with his system of symbols for the elements. Since elements combine in fixed, whole-number proportions, compounds could be shown by putting the symbols together. History (and publishers), however, favoured Jöns Jakob Berzelius's cheaper-to-print scheme that used letters.

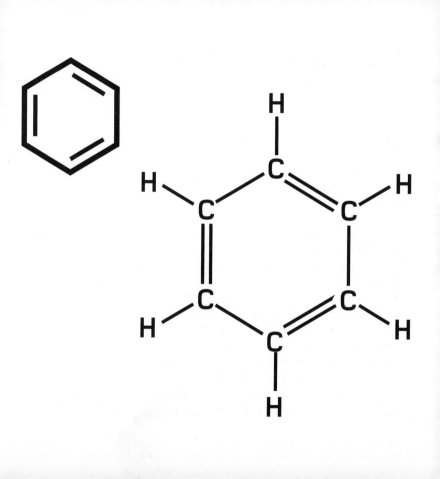

# Metals

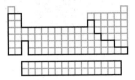

**H**ard, shiny and strong, metals are an instantly recognizable part of our world. They occupy over three-quarters of the periodic table's real estate, pushing the definitively non-metallic elements up into the top right-hand side. Ranging in reactivity from volatile to inactive (see page 38), metals are typically good conductors of electricity and heat, easily worked, ductile and dense.

Most of the bulk properties of the metal elements can be explained by their chemical bonding (see page 30). Their strong bonding gives them high melting points, although unusual quantum effects make gallium and mercury exceptions to this rule (see pages 228 and 328). The internal 'sea' of delocalized electrons transports heat and electricity easily, making metals feel cold to the touch, conduct electricity and have a surface lustre. With low electronegativity (see page 160), metals readily lose electrons to form ionic compounds with non-metal anions. They also mix with other metals in 'solid solution' alloys.

Crystalline structure
in an iron meteorite

# Reactivity series

The reactivity series is a list of the relative reactivity of metals – the ease with which metals will undergo a chemical reaction. At the top of the list are those with the highest reactivity. They will react readily and vigorously with a wide range of substances. Those with low reactivity are towards the bottom end of the list and are more choosy about the company they keep. These metals often require the encouragement of heat or high pressure to undergo a chemical reaction.

This 'league table' of reactivity is arrived at experimentally, with the reaction with water often used as a benchmark. The so-called noble metals (see page 130), such as gold and platinum, are the least reactive and will only react with strong acids. Group 1 alkali metals (see page 124) are the most reactive, and of them caesium and francium are the most violent. The reactivity series is almost identical to the ionization energy series in reverse order. This is the energy needed to remove one electron from an element (measured in the gas phase).

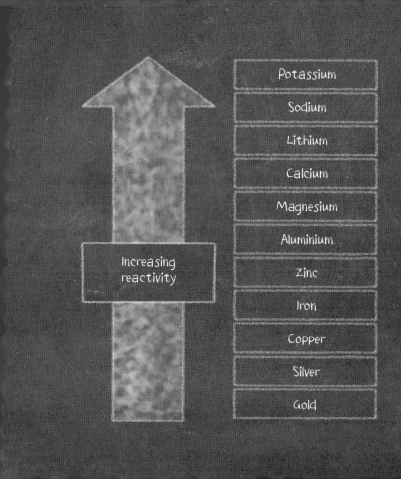

# Non-metals

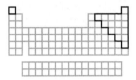

Unlike metals, which look similar and share many properties, non-metals have few uniting attributes. Instead, they are defined by their lack of metallic character. While the metal elements occupy the 'lowlands' of the standard periodic table – the left-hand side and its broad middle – non-metals are huddled in the top-right corner (with the exception of hydrogen), separated from metals by a thin line of metalloids (see page 44).

In general, non-metals are volatile – existing as gases or easily vaporized liquids and solids – and poor conductors of heat and electricity. Unlike metals, solid non-metals are dull, brittle and not malleable. They also tend to be less dense than metals, and have lower melting and boiling points (apart from carbon). With high electronegativity (see page 160) non-metal elements draw electrons from other atoms to form negatively charged anions. They form ionic compounds with metals and bond covalently with other non-metals. The noble gases (see page 150), however, are almost entirely unreactive.

NASA scientists in Antarctica fill a high-altitude research balloon with lighter-than-air helium gas.

# Allotropes

An important facet of atomic theory is that all the atoms of any given element are essentially identical, each carrying the same number of protons in their nuclei. It might seem to follow that the bulk material should always behave in the same way, but depending on how the atoms are organized, elements can take on several different forms, called allotropes, sometimes with radically different physical and chemical properties.

Allotropes of phosphorus can be red, white, black or violet. Highly reactive white phosphorus ignites spontaneously on contact with air; red phosphorus must be heated to 240°C (464°F) to burn it. These different allotropes are formed under different conditions. At great depths carbon forms diamond, a hard, non-conducting crystal. Graphite, by contrast, is a soft solid with some metallic attributes, such as a greasy sheen and electrical conductivity. While non-metals have a penchant for allotropy – particularly sulfur and carbon – the metalloids and nearly half of all common metals will also form allotropes.

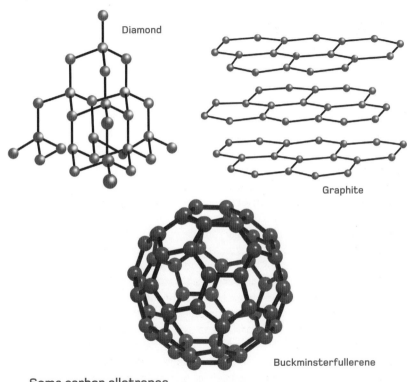

Diamond

Graphite

Buckminsterfullerene

**Some carbon allotropes**

# Metalloids

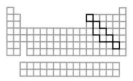

Sometimes called semimetals or poor metals, the metalloids are the 'in-between' elements with properties intermediate between those of metals and non-metals. There is a fair degree of fuzziness between metals and non-metals, reflecting nature's unwillingness to be categorized – indeed all elements have a measure of both metallic and non-metallic attributes. The metalloids, therefore, are only loosely defined. Six elements – boron, silicon, germanium, arsenic, antimony and tellurium – pick themselves automatically for the selection. These can be joined by any or all of five others – polonium, astatine, selenium, aluminium and carbon. All fall in a jagged diagonal line on the periodic table – a contested territory separating the metal elements from the non-metals. Although the metalloids often resemble metals, they exhibit the chemical behaviour of non-metals. Pure silicon is lustrous, but unlike a metal is brittle and non-malleable. Like non-metals, most will not conduct electricity, but some will under specific conditions. This allows them to be used as semiconductor materials in electronics applications.

This chunk of purified silicon comes
from an ultra-pure single crystal
some 2 metres (80 in) long.

# Nanomaterials

For many chemists, 'nano' is no big deal – after all, chemistry is the study of atoms and molecules on scales measured in nanometres (billionths of a metre) every day. However the current buzz is around nanotechnology and the microfabrication of devices on microscales. So-called nanoparticles are 10 to 1,000 times the size of an average atom: scales at which materials can behave in ways that are unusual and very different to bulk matter.

Carbon nanotubes are tiny straws made of single-atom graphene sheets rolled up on themselves. They are the strongest material known. Below scales of 50 nm, copper also becomes super-hard. Optical and electrical properties change at this size, and a vastly increased surface area to volume ratio has strange effects on solubility, diffusion and the ability to form suspensions. Among the myriad aspirations for nanomaterials are 'magic bullet' medicines, artificial photosynthesis, solar power and 'quantum dot' electronics.

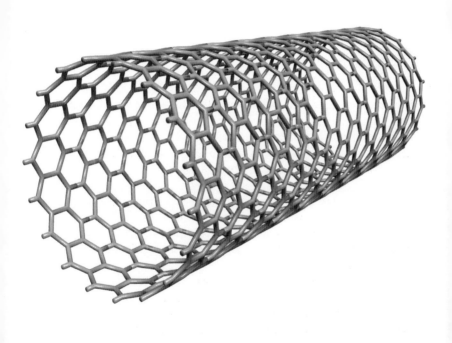

# Development of the periodic table

The question of whether there is an organizing principle behind matter in our world is an age old one. Greek philosophers, convinced that only logic and deduction could provide an insight into the harmony of the universe, attempted to analyze such things using reason alone. Meanwhile, a cadre of metallurgists, smiths, miners and alchemists learned about the behaviour of matter and, honing chemical techniques, began to build a list of 'pure', simple substances.

The success of classification systems in the 18th century, such as Linnaean taxonomy of the natural world, highlighted the need for a 'natural system of elements'. But the project to organize the chemical elements by type was frustrated by inaccuracies in measurement of atomic weights, and confusion between elements and compounds. Dmitri Mendeleev's 1869 periodic table (see page 68) swept away these inconsistencies, revealing hidden patterns (and some missing elements), and even prefiguring 20th-century discoveries of atomic structure.

# Atomic theory

The Greek philosopher Leucippus was perhaps the first to advance the idea that the world was made of tiny indivisible particles. However, his pupil Democritus (*c*.460–*c*.370 BCE) is more commonly recognized as the 'father of atomic theory'. Democritus (opposite) imagined cutting a piece of cheese in half, and then in half again and again. There must be a limit to this process, he theorized, and quickly you arrive at something that can no longer be subdivided – the very essence of cheese.

Democritus's atoms (from *atomos*, meaning 'uncuttable') had adjectival properties – solids had chunky, dense atoms, liquids were slippery, while salt and acids had sharp-edged atoms. Although this world view is by its nature untestable, its philosophical bedrock is of solid, invisible and eternal atoms. John Dalton returned to atomist principles in 1803, when he realised they could explain patterns he found among compounds (see page 60). The physical reality of atoms, however, remained unproven until the 20th century.

# The classical elements

Aristotle (384–322 BCE) roundly rejected Democritus's atomism. For this Ancient Greek philosopher, the universe had four essential poles by which everything could be categorized. All things lay somewhere between the extremes of wet and dry, and hot and cold. These related to the traditional 'elements' of earth, air, fire and water: fire was hot and dry; water cold and wet. The four qualities and four elements were sufficient to make a complete description of the terrestrial sphere. To round off the picture, Aristotle added a new element of his own invention – 'aether', the starstuff, from which the heavens are made.

Following the ideas of his teacher Plato, Aristotle held that the elements did not exist in the real world. Instead, they were idealized forms to which actual matter aspired. In this schema, the qualities become agents of change – heating cold, wet water, for example, produces hot, wet air. The behaviour of matter can also be predicted, since air and fire typically move upwards and earth and water move down.

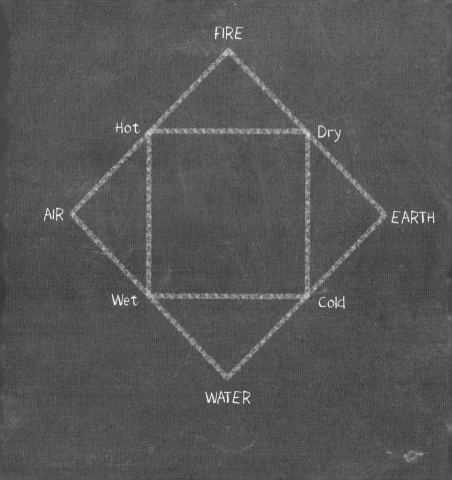

# Alchemy

Greek philosophers had little interest in practical matters, but an early brand of technical chemistry therefore emerged, almost apologetically, in the shadow of these intellectual giants. These crafts, collectively known as alchemy, focused on understanding the properties of substances and the art of manipulating them. Alchemy existed in all cultures that developed metallurgy, perhaps driven by a desire to explain how metals came to be in the ground in the first place, and by a practical interest in honing methods to refine them. Though often shrouded in mysticism and secrecy, alchemy nevertheless codified its practices and left behind a scholarly body of writing.

As such, there is no single tradition that gave rise to modern chemistry. Chinese alchemists invented gunpowder and sought elixirs to promote perfection, harmony and immortality, while Greco-Egyptian practitioners pursued the dream of gold-making. Medieval Arabic scholars Jabir ibn Hayyan (c.721–c.815 CE) and Al-Razi (854–925) pioneered many of chemistry's basic techniques.

## Chemical Signs explained.

### Acids.

1. + 🜩 vitriolic
2. + 🜩 phlogisticated
3. + 🜕 nitrous
4. + 🜕 phlogisticated
5. + 🜔 marine
6. + 🜔 dephlogisticated
7. Aqua regia
8. + of fluor
9. arsenic
10. + borax
11. + sugar
12. + tartar
13. + sorrel
14. + lemon
15. + benzoin
16. + amber
17. + sugar of milk
18. aceteous distilled
19. + milk
20. + anis
21. + fat
22. + of phosphorus
23. perlatum
24. + of prussian blue
25. aerial

### Alkalis.

26. pure fixed vegetable
27. pure fixed mineral
28. pure volatile

### Earths.

29. pure ponderous
30. pure calcareous Lime
31. pure magnesia
32. pure argillaceous
33. pure siliceous
34. water
35. vital air
36. phlogiston
37. matter of heat
38. sulphur
39. saline hepar
40. spirit of wine
41. Æther
42. essential oil
43. unctuous oil

### Metallic Calces.

44. gold
45. platina
46. silver
47. mercury
48. lead
49. copper
50. iron
51. tin
52. bismuth
53. nichle
54. arsenic
55. cobalt
56. zinc
57. antimony
58. manganese
59. siderite

A plate from Torbern Bergman's Dissertation on Elective Attractions, shows ancient alchemical symbols still very much in use in 1775.

# The Philosopher's Stone

In the classical world of Aristotle, all substances were mutable. Since a metal was an amalgam of idealized qualities, it could be transformed into another metal by adjusting the relative amounts of these internal qualities. Alchemical adherents were possessed by the idea of transmuting inferior metals into gold – a feat that could be achieved with the aid of a mystical factor called the Philosopher's Stone.

Bankrupt German merchant Hennig Brand (c.1630–c.1692) had been bewitched by the 'Great Work'. The quest for the Philosopher's Stone had swallowed his entire personal fortune when, in 1669, he isolated phosphorus. His elaborate synthesis – including boiling down 50 vats of human urine – yielded just 120 grams (4.2 oz) of white phosphorous. On contact with air, it spontaneously ignited with a blinding white light, eerily cold to the touch and suitably magical. Never realizing he was the first person to discover an element, however, Brand pressed on with his efforts to conjure up gold.

# Prout's hypothesis

By the turn of the 19th century, chemistry had left the murky waters of alchemy behind. The distinguished French 'father of chemistry', Antoine Lavoisier (1743–94), determined that mass was never created or lost in chemical reactions (as was hoped vainly by alchemists). Another French chemist, Joseph Proust (1754–1826), discovered that any given compound had the same proportions of elements by mass.

These findings fed into a growing concept of a chemical element as something unaltered during chemical reactions, changing only its arrangement and combinations as it formed compounds. In 1815, Englishman William Prout (opposite, 1785–1850) hypothesized that all elements are made from whole-number multiples of primordial, simple hydrogen atoms. Although more accurate measurement of relative atomic weights would prove him wrong, the idea was influential, and prefigures the discovery of protons. In fact, were it not for isotopes (see page 78), Prout's hypothesis would be accurate to within 1 per cent.

# Atomic weights

English chemist John Dalton (1766–1844) was the first to attempt to measure the relative masses of elements, based on his 1808 law of definite proportions. This states that when elements join to form chemical compounds, they do so in simple whole-number ratios.

To make his calculations possible, Dalton made a number of key assumptions. The 'ultimate particles' of any given element, he reasoned, must be identical. According to Lavoisier's law of conservation of mass, atoms cannot be subdivided, created, or destroyed in chemical reactions. This was the first modern atomic theory, and made the philosophically important association of atoms with elements; that elements are made up of atoms of the same type. Accompanying his theory, Dalton published a list of 30 elements alongside their masses. He also devised symbols for each element, which (despite a whiff of the esoteric) allowed formulae for compounds to be written down as the proportions of the elements they contained.

# ELEMENTS

| | | | |
|---|---|---|---|
| Hydrogen | 1 | Strontian | 46 |
| Azote | 5 | Barytes | 68 |
| Carbon | 54 | Iron | 50 |
| Oxygen | 7 | Zinc | 56 |
| Phosphorus | 9 | Copper | 56 |
| Sulphur | 13 | Lead | 90 |
| Magnesia | 20 | Silver | 190 |
| Lime | 24 | Gold | 190 |
| Soda | 28 | Platina | 190 |
| Potash | 42 | Mercury | 167 |

# Patterns in the elements

In 1818, Swedish chemist Jöns Jakob Berzelius (opposite, 1799–1848) published a list of atomic weights for most of the known elements, and a number of new elements he had discovered himself. His diligent work demolished Prout's hypothesis (see page 58), showing that the elements weren't simple multiples of hydrogen, while lending support to Dalton's atomic theory and his law of multiple proportions.

Johann Wolfgang Döbereiner (1780–1849) noticed that the weight of strontium was exactly midway between those of calcium and barium. This was the first of 'Döbereiner's Triads'. The threesome trend appeared again with the halogens chlorine, bromine and iodine, and alkali metals lithium, sodium and potassium. Linking the chemistry of the element to its physical properties was a tantalizing prospect, and foreshadowed the periodic law of chemical elements – that the chemical properties of the elements recur at predictable inervals. However, the triads remained little more than a curiosity, since many elements did not fit the pattern.

# Döbereiner's Triads

|  | Light element | Heavy element | Average weight | Middle element |
|---|---|---|---|---|
| 'Group A' | Lithium 7.0 | Potassium 39.0 | 23.0 | Sodium 23.0 |
| 'Group B' | Calcium 40.0 | Barium 137.0 | 88.5 | Strontium 87.5 |
| 'Group C' | Chlorine 35.0 | Iodine 127.0 | 81.0 | Bromine 80.0 |

Some elements with similar chemistry seem to have related atomic weights.

# Newlands' law of octaves

In 1862, the flamboyantly named French geologist Alexandre-Emile Béguyer de Chancourtois (1820–86), came up with a way to connect the numerical data of elements with their chemical properties. His *vis tellurique*, or 'telluric screw', had the elements written on a tape in order of increasing mass and wound in a spiral around a post. Elements that lined up vertically shared similar properties (every 16 places). John Newlands (1837–98) noted, however, that when ordered in this way, the properties of elements repeated every eight places (a periodicity of 7, since the noble gases were not then known), just like a musical scale. He divided the 62 known elements into 7 groups and gave each an atomic number indicating their order of increasing mass. Although it is now recognized as the origin of the periodic law of chemical elements (see page 62), Newlands was ridiculed for his 1864 'law of octaves'. His peers were unimpressed with his sometimes clumsy attempts to fit the elements to his pattern, with one snidely remarking that he might as well have organized them alphabetically.

| H | F | Cl | Co/Ni | Br | Pd | I | Pt/Ir |
|---|---|---|---|---|---|---|---|
| Li | Na | K | Cu | Rb | Ag | Cs | Tl |
| G | Mg | Ca | Zn | Sr | Cd | Ba/V | Pb |
| Bo | Al | Cr | Y | Ce/La | U | Ta | Th |
| C | Si | Ti | In | Zn | Sn | W | Hg |
| N | P | Mn | As | Di/Mo | Sb | Nb | Bi |
| O | S | Fe | Se | Ro/Ru | Te | Au | Os |

Newlands' arrangement of the elements

# Dmitri Mendeleev

Heavily bearded, bulky and bearish of temperament, the Russian chemist Dmitri Ivanovich Mendeleev (1834–1907) bestrides the history of chemistry like a colossus. This 'Einstein of chemistry' created the modern form of the periodic table – one of the icons of science. The achievement would not have been possible without his mother's grit and determination to secure her son a place at University: Maria Mendeleeva hitchhiked with the 15-year-old Dmitri across Russia, from Siberia to St Petersburg.

Deadlines can produce miracles. In 1869, running behind on the second volume of his career-defining chemisty textbook and in need of an organizing principle for the elements, Mendeleev hit on the periodic law for the chemical elements and the tabular format. His pack of 63 'element cards' accompanied him on long rail journeys as he experimented with different layouts in possibly the most fruitful game of solitaire ever played – but Mendeleev always claimed that inspiration arrived in a dream.

# Mendeleev's table

Mendeleev's periodic table at last brought order to the growing list of chemical elements. By organizing them into groups with shared chemical properties, he arrived at its unconventional pattern of short rows at the top and longer rows lower down. This matched exactly with the periodic trends (repeating patterns in physical properties) plotted by German chemist Lothar Meyer (1830–95), who independently developed a periodic law of chemical elements (see page 62) but published a year later. Where Newlands shoehorned elements into place to fit his scheme (see page 64), Mendeleev had the courage to leave blank spaces for as-yet-undiscovered elements. He even ventured to claim that certain accepted atomic weights were incorrect. The predictions of new elements with their weights, and their subsequent discoveries – *eka*-aluminium (gallium) in 1875; *eka*-boron (scandium) in 1879; and *eka*-silicon (germanium) in 1886 – caused a sensation. The periodic law was a genuine new discovery about the nature of matter, not the mere invention of a new way to organize elements.

**Ueber die Beziehungen der Eigenschaften zu den Atomgewichten der Elemente.** Von D. Mendelejeff. — Ordnet man Elemente nach zunehmenden Atomgewichten in verticale Reihen so, dass die Horizontalreihen analoge Elemente enthalten, wieder nach zunehmendem Atomgewicht geordnet, so erhält man folgende Zusammenstellung, aus der sich einige allgemeinere Folgerungen ableiten lassen.

|  |  |  |  |  |  |  |
|---|---|---|---|---|---|---|
|  |  |  |  | $Ti = 50$ | $Zr = 90$ | $? = 180$ |
|  |  |  |  | $V = 51$ | $Nb = 94$ | $Ta = 182$ |
|  |  |  |  | $Cr = 52$ | $Mo = 96$ | $W = 186$ |
|  |  |  |  | $Mn = 55$ | $Rh = 104,4$ | $Pt = 197,4$ |
|  |  |  |  | $Fe = 56$ | $Ru = 104,4$ | $Ir = 198$ |
|  |  |  | $Ni = Co = 59$ | | $Pd = 106,6$ | $Os = 199$ |
| $H = 1$ |  |  |  | $Cu = 63,4$ | $Ag = 108$ | $Hg = 200$ |
|  | $Be = 9,4$ | $Mg = 24$ |  | $Zn = 65,2$ | $Cd = 112$ | |
|  | $B = 11$ | $Al = 27,4$ |  | $? = 68$ | $Ur = 116$ | $Au = 197?$ |
|  | $C = 12$ | $Si = 28$ |  | $? = 70$ | $Sn = 118$ | |
|  | $N = 14$ | $P = 31$ |  | $As = 75$ | $Sb = 122$ | $Bi = 210?$ |
|  | $O = 16$ | $S = 32$ |  | $Se = 79,4$ | $Te = 128?$ | |
|  | $F = 19$ | $Cl = 35,5$ |  | $Br = 80$ | $J = 127$ | |
| $Li = 7$ | $Na = 23$ | $K = 39$ |  | $Rb = 85,4$ | $Cs = 133$ | $Tl = 204$ |
|  |  | $Ca = 40$ |  | $Sr = 87,6$ | $Ba = 137$ | $Pb = 207$ |
|  |  | $? = 45$ |  | $Ce = 92$ | | |
|  |  | $?Er = 56$ |  | $La = 94$ | | |
|  |  | $?Yt = 60$ |  | $Di = 95$ | | |
|  |  | $?In = 75,6$ |  | $Th = 118?$ | | |

1. Die nach der Grösse des Atomgewichts geordneten Elemente zeigen eine stufenweise Abänderung in den Eigenschaften.

2. Chemisch-analoge Elemente haben entweder übereinstimmende Atomgewichte (Pt, Ir, Os), oder letztere nehmen gleichviel zu (K, Rb, Cs).

3. Das Anordnen nach den Atomgewichten entspricht der *Werthigkeit* der Elemente und bis zu einem gewissen Grade der Verschiedenheit im chemischen Verhalten, z. B. Li, Be, B, C, N, O, F.

The world got its first sight of the periodic table in 1869, in the German periodical *Zeitschrift für Chemie*. One of Mendeleev's most important insights was that atomic weight determines the character of a chemical element.

# Discovery of the electron

At the close of the 19th century, several eminent scientists felt moved to declare that the great works of physics were complete and all that remained was increased accuracy of measurement, '...to the sixth place of decimals'. However, new discoveries in the final years of the century showed that the true nature of matter had not been adequately understood.

In 1897, J.J. Thomson (opposite) discovered that, contrary to John Dalton's solid and indestructible atoms, bits could be chipped off easily. Mysterious rays streaming from the negative terminal of his 'vacuum tube' device were made of discrete, negatively charged and exceedingly light particles. The discovery of an identical particle in beta radiation precluded the possibility that they were being supplied by electricity. The picture of matter at the smallest scale had to change radically and the result was the so-called 'plum pudding' model – negatively charged electrons distributed randomly, like fruit embedded in a stodgy 'cake' of positive charge that held the majority of the atom's mass.

# The atomic nucleus

One of the most beautiful experiments in physics is the 1911 Geiger–Marsden experiment, designed to probe the structure of the atom. Under the tutelage of the great physicist Ernest Rutherford (1871–1937), Geiger and Marsden fired alpha particles at a thin piece of gold foil. If positive charge and mass were evenly distributed across the gold atoms, as in Thomson's 'plum pudding' model (see page 70), the heavy, positively charged alpha particles could be expected to deflect only at shallow angles. However, the detectors picked up some alpha particles at angles greater than 90 degrees. Rutherford said, 'It was almost as incredible as if you fired a 15-inch shell at a piece of tissue paper and it came back and hit you.'

The gold foil experiment showed that an atom's mass is concentrated in the nucleus – matter is an unnerving 99.9999999999999 per cent empty space. To visualize this newest discovery, Rutherford postulated a 'solar system' model of the atom, with electron 'planets' orbiting a central 'star' nucleus.

# The Geiger–Marsden experiment

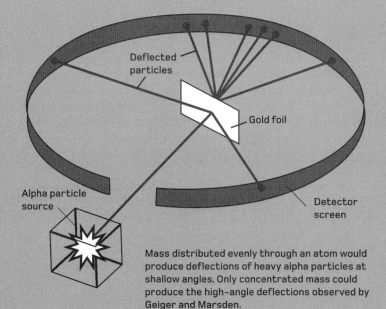

Deflected particles

Gold foil

Alpha particle source

Detector screen

Mass distributed evenly through an atom would produce deflections of heavy alpha particles at shallow angles. Only concentrated mass could produce the high-angle deflections observed by Geiger and Marsden.

# The Bohr model

Rutherford's 'solar system' model of the atom (see page 72) was never taken very seriously by scientists. It was woefully inadequate at explaining how an atom holds itself together, and how atoms join to form chemical compounds. It was known, for instance, that accelerating electric charges emit light, so an electron in a circular orbit (requiring constant acceleration towards the centre) would emit light, lose energy and spiral inexorably into the nucleus. The newly discovered small-scale structure of matter clearly required a new physics to describe it.

Aiming to explain the discontinuous absorption and emission spectra of hydrogen (see page 104), Niels Bohr (opposite, 1885–1962) suggested in 1913 that electrons occupy fixed or 'quantized' energy levels, or 'shells', around the atomic nucleus. Each level houses a certain number of electrons, after which they fill up the next shell out. Spectral lines reveal the precise energy needed for an electron to 'leap' into an outer shell, or the energy emitted when it falls back to a lower, less energetic level.

# Atomic number

In 1913, the young English physicist Henry Moseley (1887–1915) solved a riddle that had bedevilled chemists for much of the 19th century – namely, was there anything fundamental to the patterns seen in the periodic table, or was it no more than a shopping list of elements? Tables ordered by increasing atomic weight contained many niggling inconsistencies, but Moseley looked instead at the characteristic X-ray spectra emitted by elements (opposite). He found that the square root of X-ray wavelengths marked out a clear progression, implying that something fundamental was changing from element to element. An element's atomic number – its numerical order in the periodic table – had been previously considered trivial, but Moseley revealed an underlying reality to the order of elements, underpinned by a fundamental, measureable property. He speculated that the atomic number was the same as the number of protons in the nucleus, and identified gaps in the periodic table corresponding to atomic numbers 43 (now known as technetium), 61 (promethium), 72 (hafnium) and 75 (tantalum).

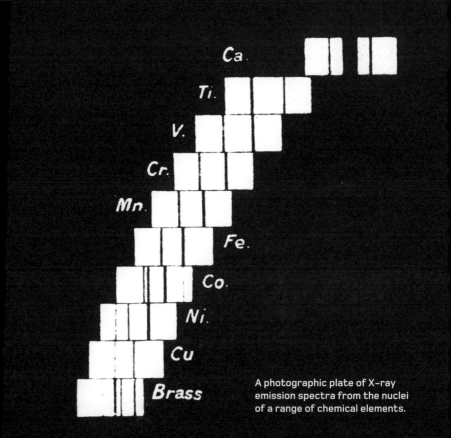

A photographic plate of X-ray emission spectra from the nuclei of a range of chemical elements.

# Isotopes

A toms have identical outsides but different insides. This is the kernel of English physicist Frederick Soddy's (1877–1956) concept of isotopes (meaning 'same place'). Isotopes are varieties of a chemical element with the same number of protons (and electrons), but different numbers of neutrons. Most elements consist of a mixture of isotopes – about one in every 6,400 hydrogen atoms, for example, has an extra neutron. This finally explains why experimental values of atomic masses are sometimes wayward and non-sequential: the weight of any sample is the average of all the atomic species it contains. In certain cases, heavy isotopes can be unstable and radioactive (see page 18).

An atom's 'mass number' is the sum of its protons and neutrons, and may differ between isotopes of the same element – hence isotopes such as carbon-12, carbon-13 and carbon-14. The modern definition of an element's 'atomic weight', meanwhile, is calculated by comparing the average mass of its atoms to that of a carbon-12 atom (defined as having an atomic weight of 12).

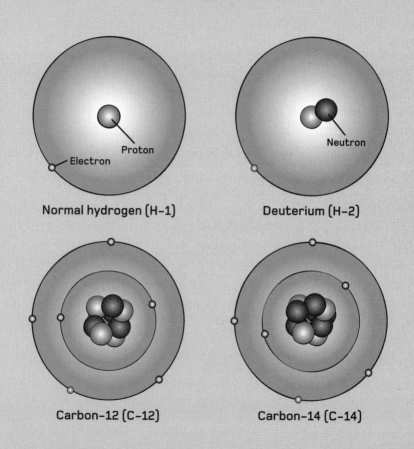

Normal hydrogen (H-1)

Deuterium (H-2)

Carbon-12 (C-12)

Carbon-14 (C-14)

# Nuclear fission

The idea of nuclear fission – splitting the atomic nucleus – was in the air in the early decades of the 20th century. Soddy and Rutherford showed that radioactive elements could spontaneously decay into another element, and in doing so, finally realized the dreams of medieval alchemists. Soddy believed that, if transmutation of elements proved possible, it would be done not for the sake of gold, but for the energy that would be released in the process. Rutherford, however, doubted that any useful energy could be extracted from the process, arguing that in practice, splitting the nucleus would take more energy than would be released. The discovery of uranium fission in 1938 won German nuclear chemist Otto Hahn (1879–1968) a Nobel Prize (although the Nobel Committee chose not to honour Lise Meitner for her part in the theoretical calculations). As well as ushering in the Atomic Age of nuclear power and nuclear weapons, it opened up a whole new arena of the periodic table, allowing the artificial synthesis of elements heavier than uranium.

# Nuclear chain reaction

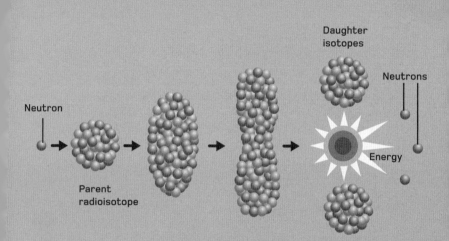

In the classical 'liquid drop model' of nuclear fission, nucleons are imagined to interact with each other like particles in a liquid. Radioisotopes can decay spontaneously, but decay can also be triggered by the impact of a neutron. In some cases stray neutrons are released as a by-product of decay, triggering a chain reaction.

# Priority disputes

The periodic table is a battlefield pockmarked with past evidence of violent disagreement. At stake for scientists were variously reputation, vindication of dearly held principles, national pride and, most tantalizing of all, the chance to write your name on some fundamental piece of the universe. Priority is key: whoever gets a discovery to press first gets the credit and the naming rights, and dirty tricks have been rife.

Personality can play a big part in such disputes. When Meyer and Mendeleev both discovered the periodic behaviour of the elements (see page 68), the German graciously gave the Russian priority. Not so French chemist Antoine Lavoisier (opposite), who claimed to have isolated oxygen independently in 1775, despite Englishman Joseph Priestley having made the discovery a year before. Priestley had even given Lavoisier his method, but the Frenchman insisted he was the first to interpret it as a new element. Swedish chemist Carl Wilhelm Scheele, meanwhile, had isolated the gas three years earlier but only published in 1777.

# Quantum physics and the periodic table

The great triumph of the periodic table – and the reason it ranks in the first order of generalizations about the universe – is its ability to accommodate new scientific discoveries. Its first period has two elements – hydrogen and helium. The next two have eight, while the next period is much longer, with 18 elements running horizontally. Historically, this 2, 8, 8, 18… sequence was simply what fell out when the elements were ordered by atomic number and grouped by shared chemistry, but the advent of quantum physics revealed its deeper significance.

Electrons within the atom (see page 10) are confined to fixed, quantized energy levels, each of which can house a certain number (revealed by the emission and absorption spectra of elements – see page 104). This pattern matches the order of the periods, with the first shell, or energy level, holding two electrons, the next eight and so on. The order in which electron orbitals within the shells fill up also defines another feature of the modern periodic table – the valence blocks (see page 122).

Gold nanoparticles have been used by artists for centuries. The vibrant colors they produce when scattered through materials such as glass are due to the way light interacts with electrons in atomic orbitals.

# The chemistry of elements

Alchemists, chemists and physicists have strived to understand the nature of matter for millennia: where do elements come from, how are they distributed and what governs their chemistry? Our best answers to the first and second questions are that the lightest elements came into being shortly after the Big Bang (see page 88), while midweight elements are forged in stars and the heaviest are spread across space by supernova explosions (see page 90). The relative abundances of elements in the universe reflect this order, although their distribution in our solar system is affected by the events of planetary formation (see page 92). As to the third question, the chemical behaviour of elements is now understood in terms of their electron configuration (see page 14).

In 1789, French chemist Antoine Lavoisier formulated the modern definition of an element as a substance that cannot be chemically or physically simplified. In doing so, he laid an experimental pathway for isolating elements, leading eventually to new discoveries beyond the realm of naturally occurring elements (see page 114).

Although oxygen had been isolated before, Antoine Lavoisier's interpretation of the substance as an element was a watershed moment, marking the beginning of modern chemistry.

# Cosmic abundance
# of elements

The outpouring of energy that gave birth to the universe – the so-called Big Bang – produced a riot of subatomic particles. After about ten seconds, once they had cooled sufficiently, mutually attracted particles could clump together to create the nuclei of hydrogen atoms (see page 170). Some of these then fused together to form helium nuclei and a tiny amount of lithium. The work of about 20 minutes remains more than 13.8 billion years later – atoms are pretty much immortal, and much of this early hydrogen and helium is still with us, and the primordial ratio of elements in the universe remains at 74 per cent hydrogen, 24 per cent helium and 1 per cent everything else. For this reason, astronomers categorize the entire 118 known elements into hydrogen, helium and 'metals'. Only the most massive stars fuse elements heavier than helium (see page 90): during this process, elements with odd numbers of protons are more likely to capture an additional proton than those with even numbers, making elements with even atomic numbers considerably more abundant than those with odd numbers.

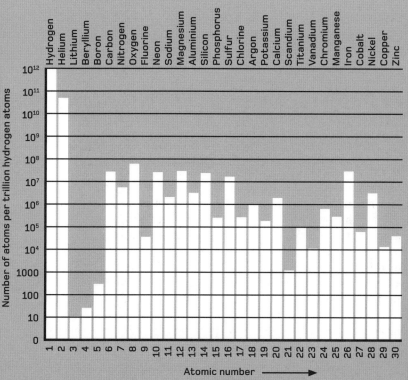

Cosmic abundance of lighter elements

# Stellar nucleosynthesis

The theory of stellar nucleosynthesis explains the origins of the chemical elements and their relative abundances. A paper published in 1957, affectionately titled $B^2FH$ (authored by Geoffrey Burbidge, Margaret Burbidge, William Fowler and Fred Hoyle), first revealed how they are fused in the cores of stars.

Deep inside stars, hydrogen nuclei join together to form helium. The mass of a helium atom is infinitesimally smaller than the combined mass of the hydrogen nuclei that go into it, and this excess mass is released as energy. As they run out of hydrogen in their cores, more massive stars fuse helium and successively heavier elements, creating heavier elements up to iron (atomic number 26). Eventually, however, even the most massive stars run out of fuel they can viably fuse. In such stellar monsters, the core collapses suddenly and a supernova shockwave tears the star apart, producing high temperatures and pressures that briefly allow the heavier elements of the periodic table to be created and strewn across interstellar space.

A supernova scatters most of a
dying star's material across space.
In the heat and fury of the explosion
elements heavier than
iron are fused.

# Elements in the Earth

The situation on Earth could not be more different than space. The three most abundant elements on our planet's surface are oxygen, silicon and aluminium. They combine to form the stable silicate minerals that make up the crust. Earth's most common element, iron, is only the fourth most abundant in the crust, for reasons that date back some 4.6 billion years.

The solar system formed from a large cloud of gas and dust that fell in on itself under its own gravity. When the Sun was born at its centre, the young star's intense radiation drove light volatile gases away to the far reaches of the solar system: this is why the planets close to the Sun are rocky, while those farthest away form the gas giant planets. Energy released during Earth's formation was enough to melt the entire planet: heavy elements such as iron sank to the core, and the so-called 'siderophiles' (iron-loving metals such as nickel) went along for the ride. Scientists think that the unique balance of elements on Earth played a crucial role in the genesis of life.

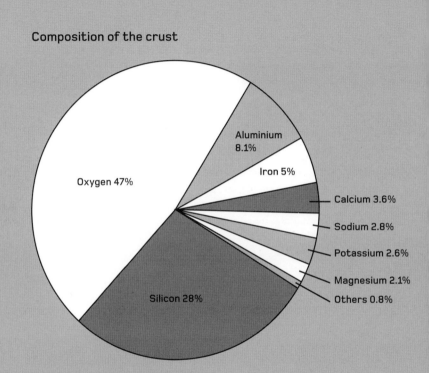

# Composition of the crust

Oxygen 47%

Silicon 28%

Aluminium 8.1%

Iron 5%

Calcium 3.6%

Sodium 2.8%

Potassium 2.6%

Magnesium 2.1%

Others 0.8%

The separation of our planet into layers of different densities early in its history left the Earth's crust depleted of iron and associated metals.

# Native element minerals

Only 13 of the substances we recognize as elements were known to the ancients. These 'native elements' are found unalloyed with any other substance, in the form of minerals (naturally occuring inorganic substances). Liquid mercury can ooze out of another mineral, cinnabar, while sulfur forms pure crystals around volcanic fumaroles. Other non-metals found in mineral form include the near-twins arsenic and antimony. Carbon forms crystalline diamond at depths of 140–190 kilometres (85–120 miles), or graphite closer to Earth's surface.

Several other metals are found uncombined – copper, zinc, silver, gold, tin, bismuth and lead – but Earth's powerful oxidizing atmosphere means that most metals end up in oxide or sulfide compounds. Native tin is exceptionally rare, while the only unoxidized native iron on the planet arrives from outer space within iron–nickel meteorites. Naturally occurring alloys include electrum (a mixture of gold and silver) and platina (an indeterminate mix of platinum-family metals).

Sulfur deposits around a volcanic fumarole

# Smelting

Most metal elements are not found in their native state, but are locked into minerals within rocks. Ores are those minerals that contain a high proportion of metal, are readily available and from which extracting metals is economical. Most are oxide or sulfide compounds, and the process of extraction, or 'smelting', uses a chemical reaction to separate them.

The first stage is roasting, to drive off any sulfur and carbon and leave an oxide. Then, the oxygen is removed by heating with carbon, usually in the form of coke or charcoal. The carbon combines with oxygen to form carbon dioxide, escaping the furnace as gas. Adding a flux, such as fluorite, lowers the temperature of the slag (waste) and makes it less viscous and easier to drive off. Tin and lead were the first metal elements to be smelted – cast-lead beads from Anatolia, Turkey, have been found dating to around 6500 BC. Today, metals are often concentrated before smelting and purified afterwards – modern electrolytic refining achieves 99.99 per cent pure copper.

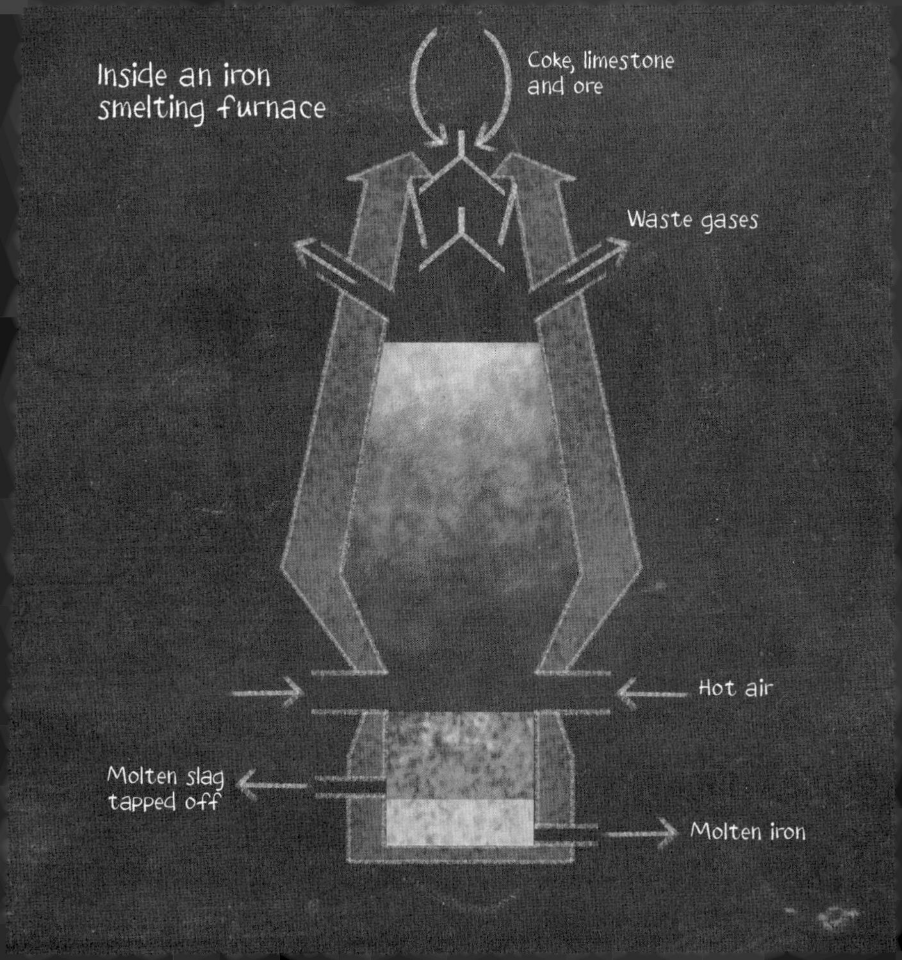

Inside an iron smelting furnace

Coke, limestone and ore

Waste gases

Hot air

Molten slag tapped off

Molten iron

# Discovery of airs

The discovery of invisible gases surrounding us was one of those perceptual leaps that periodically occur in science and require us to completely recast our ideas about the world. Jan Baptist von Helmont (1579–1644), who coined the word 'gas', speculated that the mass discrepancy between wood and the ash left after it had been burned, was made up by some insubstantial airy substance lost to the surroundings. Aristotle had allowed only one gaseous element – air – but we now recognize 11 elements that are stable gases under standard conditions. The first of these 'airs', hydrogen, was isolated in 1766 by Henry Cavendish (1731–1810). One of the so-called 'pneumatic chemists', he pioneered the technique of reacting metals with acids and collecting the gases released over water or mercury. Cavendish also showed that normal air has several different components, but interpreted them in terms of 'phlostigon' – a fire-like substance that promoted combustion. Antoine Lavoisier's 'chemical revolution' (see page 86) came as a result of recognizing oxygen as an element.

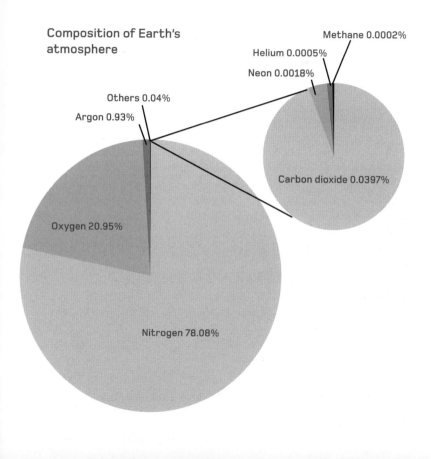

Composition of Earth's atmosphere

Methane 0.0002%

Helium 0.0005%

Neon 0.0018%

Others 0.04%

Argon 0.93%

Carbon dioxide 0.0397%

Oxygen 20.95%

Nitrogen 78.08%

# Reduction of oxides

For 18th-century 'apothecary-chemists', the tried and tested method for separating mineral ores into their component parts was reduction of an oxide, or 'earth'. To reach the high temperatures that many metals require to form oxides required delicate glass blowpipes, which remained a standard part of a mineral 'assaying' kit up to the development of X-ray analysis of minerals in the 20th century. Since most properties could be determined from the oxide, often no attempt would be made to isolate an element.

Contemporary giants whose names have since faded included Andreas Marggaf (1709–82, discoverer of zinc), Carl Wilhelm Scheele (1742–86; molybdenum, tungsten, barium, hydrogen and chlorine) and Martin Heinrich Klaproth (1743–1817; zirconium, uranium and titanium). They and others advanced the sophistication of techniques used to analyze compounds in minerals – in particular through gravimetric analysis, the careful weighing of reactants and products of a chemical reaction.

Laboratory of the English apothecary-chemist William Lewis, *c.*1760

# Electrolysis

The most fruitful decade of all for element discoveries was 1800–10, when the strongly bound compounds of the group 1 and 2 metals yielded to the newly discovered electromagnetic force. The man wielding that power most effectively was Humphry Davy (1778–1829). Using Alessandro Volta's newly invented battery, the English chemist pummelled metal oxides with electrical energy until they split into their component elements. Davy isolated no fewer than six new elements using this technique – potassium, sodium, barium, boron, calcium and magnesium.

The process known as electrolysis involves passing a direct current through a molten ionic substance or a dissolved ionic salt. The charged ions within the liquid carry the electricity, causing them to migrate to the oppositely charged terminals. This separates the compound, and the component elements can then be collected at each electrode – a chemical reaction that would not happen spontaneously. Jöns Jakob Berzelius (opposite) used the process to develop a theory of electrochemistry.

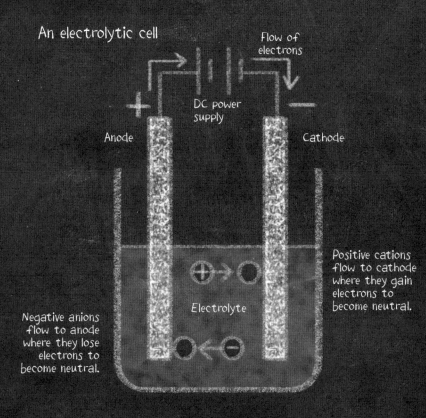

# Spectroscopy

**D**eveloped in 1860 by Robert Bunsen (1811–99) and Gustav Kirchhoff (1824–87), spectroscopy allowed elements to be identified through characteristic frequencies of emitted or absorbed light. When matter is heated, electrons absorb energy to gain an excited state, and then re-emit an identical quantity of energy as electromagnetic radiation as they fall back to the 'ground state'. The frequencies absorbed (and emitted) depend on the electronic configuration of the atoms, and so each substance has its own specific absorption and emission spectra. Bunsen and Kirchhoff crowned their discovery by finding two new elements – caesium (named for its distinctive sky-blue spectral lines) and rubidium (with characteristic red lines). Ultimately, the technique enabled the discovery of 15 new rare earth elements and called into question certain historical 'discoveries'. For the first time, we could analyze light from objects in space to see what they were made of, and this resulted in the discovery of an important new element – helium – in the Sun (see page 172).

A highly dispersed spectrum of light from
the Sun reveals countless absorption lines
caused by gases in its atmosphere.

# Distillation of liquid air

The method for separating the different parts of air is called fractional distillation. A fractionating column relies on the differences in volatilities (boiling points) of component parts in a liquid mixture. In the laboratory, the liquid is heated in a flask and allowed to rise up a column packed with glass beads or metal rings. As the vapour rises, it condenses on the beads and vaporizes again, often many times over. Each time, the gas becomes enriched in volatile components, until the 'fraction' exiting at the top of the column contains only one substance. The operation is repeated with increasing temperature, driving off successively less volatile fractions each time.

Eighteenth-century pneumatic chemists came agonizingly close to isolating argon in this way: several reported an 'unreactive portion' in the air, or a nitrogen component of slightly higher density. By liquefying air and then carefully distilling it, Scottish chemist William Ramsay (1852–1916) isolated argon in 1892 and eventually found an entire new group of elements (see page 150).

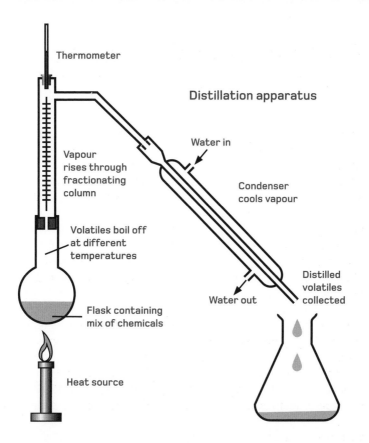

**Distillation apparatus**

Thermometer

Vapour rises through fractionating column

Volatiles boil off at different temperatures

Flask containing mix of chemicals

Heat source

Water in

Condenser cools vapour

Water out

Distilled volatiles collected

# Carbon chemistry

With more than 10 million known carbon-based molecules, the sheer variety of carbon compounds is bewildering. They range from hydrocarbon fuels, alcohols and fragrant esters to plastic polymers, toxic benzene rings and high-tech graphene. They encompass essential biomolecules, such as amino acids and proteins, not to mention DNA and ATP, life's twin pillars of form and energy. It's little wonder that 'organic' chemistry – devoted to studying these compounds – is an entire separate discipline.

Until 1828, it was almost taken for granted that organic chemicals carried some inherent 'vital force'. Friedrich Wöhler (1800–82) changed all that, however, when he synthesized urea from inorganic components. Since then, studying the reactions of organic chemicals and their interactions inside cells has helped us to understand the workings of living systems, mimic natural drugs and design new pharmaceuticals. Industrial synthesis of carbon compounds such as dyes, meanwhile, led to the development of the petrochemical industry.

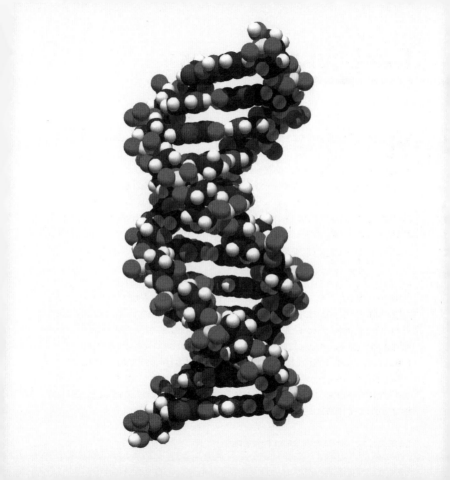

# Isolating aluminium

Some elements are bound very tightly to their minerals – some 17 'rare earth elements' took more than a century to isolate (from the discovery of yttrium in 1794, to promethium in 1947). Aluminium, too, proved difficult to separate from its ores: identified by Humphry Davy in 1808, it wasn't until 1827 that Friedrich Wöhler isolated it as a metal. Even in the 1880s, the scarcity of pure aluminium meant it commanded a higher price than gold, despite being the third most abundant element in Earth's crust (see page 92).

Then in 1886, French chemist Paul Héroult and American chemist Charles Hall torpedoed aluminium commodity prices with their independent discovery that dissolving aluminium salts in a different electrolyte (see page 102) would permit easy electrolysis at last. Electrolysis of alumina (aluminium oxide) in molten cryolite (a sodium aluminium fluoride mineral) resulted in pure molten aluminium, and this 'Hall–Héroult process' (opposite) is still the primary method for extracting aluminium metal today.

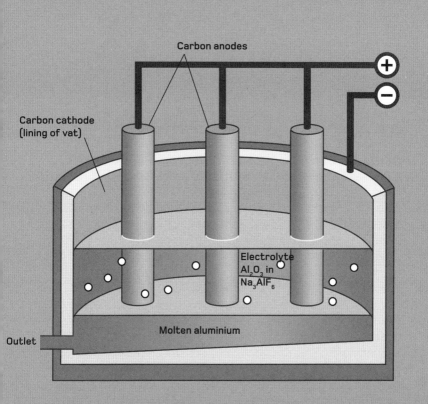

# Discovery of radioactive metals

The techniques that permitted the first isolation of radioactive elements involved phenomenal patience and colossal endurance. Working in a draughty, leaky outhouse, Pierre and Marie Curie separated the components of radioactive pitchblende ore from a boiled 'soup', relying on the fact that different compounds crystallize out of the hot mixture at different rates. Each tonne of ore yielded just 0.1 g of radium chloride, but the discoveries made by French physicist Pierre, his Polish émigré wife Marie and, later, their daughter Irène, changed our understanding of matter itself.

Marie Curie's realization that the energies coming from her radioactive pitchblende and chalcocite ores were coming from within individual atoms meant that atoms themselves could not be indivisible, while the discovery that elements changed their identity during this decay meant that they were not immutable. Indeed, the transmutation of elements happens naturally, and can even be used to calculate the age of our planet.

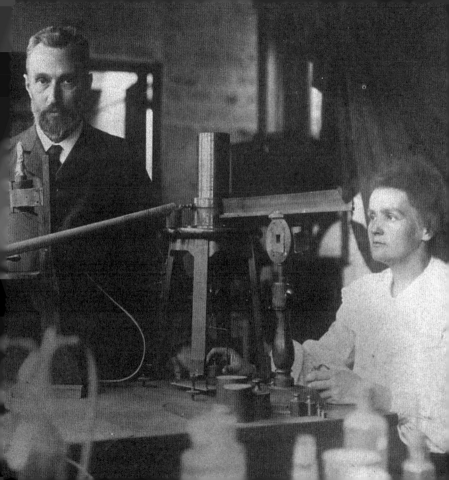

# Making new elements

James Chadwick's (1891–1974) discovery of the neutron in 1932 at last made the dreams of alchemists a reality. Enrico Fermi (1901–54) saw that it was possible to add mass to an atomic nucleus by firing neutrons into it. Since bulky nuclei are prone to radioactive decay that changes both mass and atomic number, this opened up the possibility of creating new elements.

Techniques pioneered by Glenn T. Seaborg (1912–99) and Albert Ghiorso (1915–2010) in the late 1940s used neutron bombardment in nuclear reactors to produce elements heavier than plutonium, and add new real estate to the periodic table. To create 'superheavy' elements, however, more energy was needed. Ernest Lawrence's (1901–58) 1.5-metre (60-in) cyclotron at Berkeley, USA, whipped alpha particles close to the speed of light before slamming them into a target, and produced several new elements. The quest for ever heavier elements is now an international effort, with superheavy ion research facilities at Lawrence Livermore in the USA, JINR in Russia, GSI in Germany and RIKEN in Japan.

Glenn T. Seaborg (left) and Edwin M. McMillan (right) with the 60-inch cyclotron, which they used to discover plutonium, neptunium and many other transuranic elements.

# Imaging atoms

For most of its scientific existence, the atom has been a conceptual tool for chemists and physicists – useful for explaining the reaction mechanics, conservation of mass and numerous interactions of energy and matter. In 1981, however, this layer of abstraction was swept away by the ability to *see* atoms. Gerd Binnig (b.1947) and Heinrich Rohrer (1933–2013) produced the first images of silicon atoms using their invention, the scanning tunnelling electron microscope (STM). Astoundingly, they were also able to manipulate single atoms, nudging a few iron atoms into a circular 'corral' on a copper surface.

The STM was the first instrument in the field of atomic force microscopy. It uses a nanoscale 'probe' that responds to the 'roughness' on a surface. Sensitive to resolutions not available to light microscopy, it allows variations in force to be plotted as a map that reveals individual atoms and the interactions between them. In 2013, researchers using an STM made the first observations of chemical bonds forming in real time.

# Patterns in the periodic table

Originally laid out using shared chemical behaviour as an organizing principle, the periodic table gives rise to 'chemical families' of related elements. Metals account for over 75 per cent of elements, falling into alkali, alkaline earth, transition, lanthanoid, actinoid and – with less strongly metallic characteristics – post-transition metals (see pages 124, 126, 128, 134, 152 and 138 respectively). A thin strip of metalloids (see page 44) separate the metals from the non-metals (see page 40), while the noble gases run down the extreme right-hand side of the table (see page 150).

More than a simple spreadsheet for the elements, the periodic table is a philosophical construct, which has been robust enough to incorporate later discoveries of atomic number (see page 76), electron configuration (see page 14) and isotopes (see page 78). Each entry represents a set of isotopes, a kind of 'median' that Dmitri Mendeleev called the 'real element'. The table is a map to the building blocks of matter, revealing at a glance an element's make-up, behaviour and the direction in which properties change.

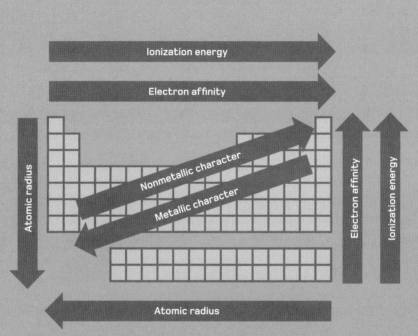

The power of the periodic table is not only how it arranges its 105 solids, 11 gases and 2 liquids, but how it reveals the hidden relationships and patterns between the elements.

# Groups and periods

The periodic table is laid out with groups forming vertical columns, and periods in horizontal rows. Schoolchildren are commonly taught that the main body elements – those in groups headed by lithium, beryllium, boron, carbon, nitrogen, oxygen, fluorine and helium – are numbered groups 1–8. However, this system gets messy when it's necessary to talk about the 10 columns of transition elements lying between groups 2 and 3 (see page 128). IUPAC – the International Union of Pure and Applied Chemistry – endorses a 1–18 nomenclature to avoid complication. The lanthanoid and actinoid series (see pages 134 and 152) fall between groups 2 and 3. Periodic table rows run from period 1, with only 2 elements, through to period 7, with 32 elements. Atomic number increases along a period, in increments of one – equivalent to adding a proton to the atomic nucleus each time. The different number of elements in each period corresponds to the maximum number of electrons held by each successive energy level. Elements in a group have identical valence shell electron configurations (see page 14).

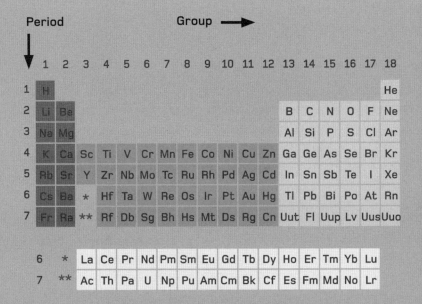

# Blocks

As well as its groups and periods, the periodic table has a 'hidden feature', with elements falling into zones called blocks. Corresponding to the broad groupings of 'types' – alkali and alkaline earth metals, transition metals and non-metals – these regions have a deeper underlying significance, based on the electron configuration of the elements they contain.

The blocks are named after the orbital in which the highest-energy electron resides. There are four in all: the s-block (groups 1 and 2) have their outermost electrons in s-orbitals; the p-block (groups 13–18) have their valence electrons in the p-orbitals; d-orbital electrons define the d-block (groups 3–12); and the f-block elements are the lanthanoid and actinoid series with electrons occupying f-orbitals. Helium is an anomaly: it has both its valence electrons in the s-orbital (a full first shell), but sits with the p-block elements of group 18. The position of the blocks reflects the order in which electron orbitals fill up (see page 14).

# Blocks of the periodic table

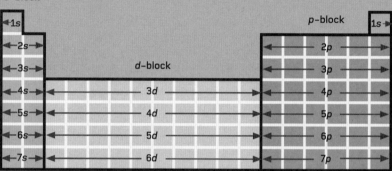

# Alkali metals

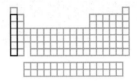

**G**roup 1, on the extreme left of the periodic table, is occupied by soft, low-density and reactive metals. Most are light enough to float on water and show their shininess only momentarily on freshly cut surfaces (since they react readily with air to form a tarnishing oxide layer). These metals are never found uncombined in nature and must be kept under oil or in an inert atmosphere to stop them reacting.

The alkali metals are a close-knit bunch of elements, sharing a greater family resemblance than many other groups on the periodic table. Their reactivity stems from the lone electron in their valence shell, which is enthusiastically discarded to give a stable full outer shell. These metals rip water molecules apart, releasing hydrogen gas and leaving a strongly alkaline metal hydroxide (hence the group name). The violence of their reactions increases down the group: lithium fizzes when added to water, but potassium explodes with purple flames. Caesium is the most reactive metal element on the periodic table.

Potassium undergoes a
violent reaction when it is
dropped into water.

# Alkaline earth metals

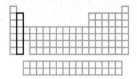

Beryllium, magnesium, calcium, strontium, barium and radium are another group of low-density, reactive metals with low melting and boiling points. Shiny and soft like the alkali metals, these group 2 elements appear rarely in their native form. They form stable oxides (known in antiquity as 'earths') that are important components of the Earth's crust. Like their feisty neighbours in group 1, the group 2 metals react with water, but not quite as explosively: the one extra proton in the nucleus pulls in electrons in the s-orbital just a little tighter, requiring a little more energy to remove them, and reducing reactivity somewhat. However, as cations with a +2 charge, they are highly reactive.

Beryllium is a bit of an outsider among the alkaline earth metals. It is hard, exceptionally stiff and has a rather high melting point. Magnesium and calcium play important roles in Earth chemistry and in the body, as does strontium. Radium holds a special place in the periodic table as the first radioactive element, discovered in 1898 (see page 344).

Transition metals such as nickel are, on the whole, much less reactive than alkali or akaline earth metals.

# Noble metals

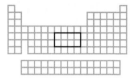

Within the broad group of transition metals, there are smaller sets of elements that share similar properties. The 'noble' metals are unreactive – echoing the inert noble gases of group 18 (see page 150) – and resistant to corrosion. Unlike the so-called 'base metals', they do not oxidize in moist air. The noble metals are ruthenium, rhodium, palladium, silver, osmium, iridium, platinum and gold. To this family are sometimes added mercury, rhenium and copper.

These elements are among the densest and rarest metals. Because of their scarcity and resistance to corrosion, they are valuable, and include the precious metals that are used for jewellery and as capital assets. The noble metals also encompass the highly prized platinum group metals (PGMs) ruthenium, rhodium, palladium, osmium, iridium and platinum. These elements have numerous applications and are particularly useful as catalysts in industrial processes. Their imperviousness makes them the material of choice for ultra-high vacuum equipment.

Once thought to be 'immature' gold, today platinum is a symbol of eternity.

# Coinage metals

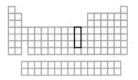

The elements of group 11 – copper, silver and gold – are often referred to as the coinage metals. This is a group defined more by cultural and historical significance than by chemical and physical properties. To stay in circulation for 30 years as currency, materials should be unreactive, malleable (not brittle) and hard-wearing. Metals are the obvious choice.

Historically, currency was minted using high-value metals. More recently there has been a shift towards less valuable elements to avoid the problems encountered when the value of a metal in a coin exceeds its face value. 'Copper' coins no longer contain a high proportion of copper – US pennies are made of copper-clad zinc and British pennies are copper-plated steel. Despite its name, the famous US 'nickel' only contains 25 per cent nickel – but even this cupronickel alloy is being phased out as the price of nickel rises. Intrinsically valuable metals, such as gold and silver, are now mostly used as a store of value in official reserves or private bank vaults.

# Lanthanoids

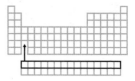

First discovered near the Swedish town of Ytterby in 1787, the lanthanoids (aka lanthanides) are a series of 15 metals with strikingly similar properties. They were first identified by Finnish chemist Johan Gadolin (1760–1852) in the mineral ytterbite (later renamed gadolinite). A full eight elements have names related to Ytterby, Gadolin or Scandinavia in general – yttrium, ytterbium, terbium, erbium, gadolinium, scandium, holmium and thulium.

The lanthanoids lie in period 6 on the periodic table, but are one of two additional rows usually shown beneath the main table – the full-length 'expanded' table with the lanthanoids in their proper place is deemed too unwieldy for general use. They are so-called f-block elements (see page 122) as their valence electrons fill up the outer f-orbital. When cut, they have a silvery shine, but tarnish quickly in moist air. Rare earth elements are a group of economically important metals including the lanthanoids, scandium and yttrium – their alloys make powerful permanent magnets used in wind turbines, MRI machines, mobile phones and computers.

The disused quarry at Ytterby, Sweden is now designated a historic landmark.

# Post-transition metals

The post-transition metals are those elements that do not have the incomplete d-orbitals of the transition metals but nevertheless share some of the same properties. They are located on the periodic table between the transition metals and metalloids, to the right of the d-block (see page 122). There is no single accepted definition of a post-transition metal, so these 'borderline' elements rather depend on where the lines separating transition metals and metalloids are drawn. Traditional post-transition metals are gallium, indium, thallium, tin, lead and bismuth, but the set can be expanded to include the group 12 elements, zinc, cadmium and mercury, along with aluminium, germanium, antimony and polonium. The chemistry of the radioactive period 7 elements (copernicium, ununtrium, flerovium, ununpentium, livermorium, ununseptium and astatine) may also be 'post-transitional'. While the post-transition elements have metallic character, they tend to be softer, less strong, less dense and less able to conduct heat and electricity. They also have lower melting and boiling points.

Bismuth is a dense metal with a low melting point and a tendency to form beautiful stepped crystals.

# Toxic heavy metals

A heavy metal is any high-density metal or metalloid that is toxic to human beings in low concentrations. Lead, mercury, cadmium and arsenic are perhaps the most notorious of these environmental pollutants. Although many heavy metals are rare in the earth, they are concentrated to toxic levels by mining and industry, and the waste these activities produce.

Mining operations often use heavy metals to extract precious metals. They become concentrated in tailings and containment ponds, and leach into soils or escape as run-off. Coal also contains heavy metals, which are released into the air when it is burned or concentrated in soot. Batteries are sources of lead, cadmium and nickel, while natural arsenic contamination of groundwater is a major problem in parts of Asia, South America and the USA. Heavy metals interfere with body chemistry, attaching to biomolecules and stopping them working properly. They accumulate in the tissues of plants and animals, building up to dangerous levels in the food chain.

# The boron group

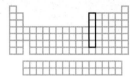

**B**oron is the odd one out in its own family of elements. The others – aluminium, gallium, indium and thallium – are generally soft, silvery, weak metals. Boron, in contrast, is a hard metalloid. Where boron is largely unreactive, the other members of group 13 readily combine with other elements. The boron family of elements is commonly found in ores and minerals – aluminium is the third most abundant element in the Earth's crust. The group all have three electrons in their outer shell, and sometime go by the name 'triels' or 'icosagens'. The metallic character of the elements and their reactivity increases down the group. So, while indium and thallium are true metals, aluminium and gallium will form both ionic and covalent bonds. This trend, which is also seen in groups 14, 15 and 16, can be explained by the increasing size of the atoms in the group: boron's outer electrons are held tightly to the nucleus, decreasing its size and reactivity, but the valence electrons of heavier atoms are shielded from the nucleus by inner electron shells, making them larger and more reactive.

Crystals of alum
(potassium aluminium sulfate)

# The carbon group

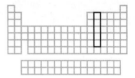

The carbon family, also known as the crystallogen elements, is a collection of individuals rather than a team. The single thing that they have in common is four valence electrons – hence they are sometimes called the tetrels or tetragens.

The metallic attributes of the group 14 elements – shiny lustre, malleability, heat and electrical conductivity, for example – increase down the group. Carbon, at the head, is a non-metal, while silicon and germanium are metalloids with properties intermediate between metals and non-metals. Tin and lead are properly metallic. Accordingly, these heavier members of the carbon group tend to form cations (with anything up to a +4 oxidation state) and join to non-metal elements with ionic bonds, or release their valence electrons to take part in metallic bonding. Carbon and silicon instead favour covalent bonding, and carbon in particular – the crucial element for life – has such a diversity of bonds available to it that an entire branch of 'organic' chemistry is dedicated to its study (see page 108).

A raw diamond against the diamond-bearing mineral kimberlite.

# The nitrogen group

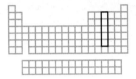

The group 15 elements are sometimes – but not often – called the 'pnictogens'. This strange-sounding appellation refers to the 'choking' or 'stifling' properties of nitrogen. An inert gas, nitrogen will effectively smother and extinguish a flame. Phosphorus, in contrast, is anything but chemically inactive – white phosphorus ignites spontaneously on contact with air and is a terrible weapon of war, while matchtips are coated with red phosphorous.

The nitrogen group have deep and arcane alchemical associations. Both bismuth and phosphorus were discovered by alchemists (see page 56). Arsenic was isolated from the ancient minerals realgar and orpiment, while stibnite – antimony ore – was an essential component of 'Greek fire', an early incendiary weapon. Like many of the groups in this region of the periodic table, the pnictogens change character, with two non-metals (one gas and one solid), two metalloids and one metal. With five valence electrons, a variety of bonding is available to them.

About 85 per cent of the body's phosphorus is contained in the bones.

# The oxygen group

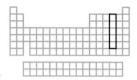

**G**roup 16 are the 'ore formers' or chalcogens (pronounced with a hard 'ch-', like chemistry). Mostly non-metallic, this group contains the plentiful elements oxygen and sulfur. Between them, these two lock up many of Earth's metal resources into the stable oxide and sulfide minerals that comprise the world's most important metal ores. Oxygen is by far the most abundant element in the crust, and any substance exposed to Earth's atmosphere or water comes instantly under attack from this reactive, electronegative element. Most tarnish rapidly, developing an oxide coating.

Tellurium (opposite) is a metalloid, selenium is variously considered both a non-metal and a metalloid and the status of polonium is unclear. These varied properties are explained by electron configuration: with six electrons in its outer shell, a chalcogen can gain two electrons to form an anion with a charge of -2 and fill the shell. However, it can also lose four electrons to leave two electrons in its s-orbital, or even shed all six.

Tellurium is an untypical
element in a varied and unusual
group of chemical elements.

# Halogens

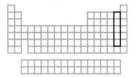

The penultimate group of the periodic table marks a return to the uniform and unified categories of elements seen in the early groups. The halogen elements of group 17 share remarkably similar properties, forming an unruly bunch of reactive non-metals. Just one electron shy of a complete outer shell, it seems as if they are willing to employ any means to obtain one. This electron-stealing behaviour is known as electronegativity, and the most electronegative element of all is fluorine. Because of this power (and the willingness of metals to donate electrons), the group 17 elements mostly act as the negative charge-bearing partner in ionic compounds (see page 26).

Fluorine and chlorine are gases, bromine is one of only two liquids on the periodic table at standard temperature and pressure, while iodine and astatine are solids. Astatine is among the rarest radioactive elements of all, while the superheavy halogen ununpentium has so far only been reported experimentally in particle accelerators.

In contrast to the uncombined elements, halogen compounds are often exceptionally unreactive. They include PTFE (Teflon) and flame-retardant organo-bromines.

# Noble gases

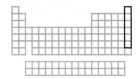

Group 18 is an anomaly. Most groups of the periodic table contain no gases at standard atmospheric conditions, and of those that do, only the lightest one or two members are gases. The noble gases – helium, neon, argon, xenon and radon – by contrast, are *all* gaseous. Argon makes up a small, but significant part of the Earth's atmosphere. Radon, the heaviest noble gas, is highly radioactive. The discovery of this entire set of new elements – mostly by Scottish chemist William Ramsay (see page 106) – was a huge scientific breakthrough. Mendeleev, having missed predicting the group's existence, could not quite believe it (even though it bolstered his periodic law). At the far edge of the periodic table, the noble gases are characterized by extreme unreactivity due to the stability afforded by complete electron shells. They are 'ideal' gases with very little attraction between their monatomic particles, and consequently low melting and boiling points (see page 158). Although noble gases were once thought to be entirely inert, in 1962 British chemist Neil Bartlett succeeded in synthesizing xenon hexafluoroplatinate.

# Actinoids

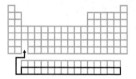

The actinoids – also known as actinides – are the second row of 'f-block' elements (see page 122). Found on period 7, they are normally displayed for convenience underneath the main body of the periodic table, along with the lanthanoid elements (see page 134). Their relationship to the rest of the elements – lying between groups 2 and 3 –becomes clearer when they are included in an otherwise-ungainly 'wide format' table.

Running from actinium (89) to lawrencium (103), these heavy elements are all radioactive. Four – the eponymous actinium, thorium, protactinium and uranium – occur naturally on Earth. The half-lives of thorium and uranium (the time taken for half of a sample to decay through radioactivity) are counted in billions of years, so they are found in appreciable quantities. The so-called 'minor actinoids' exist in vanishingly small amounts, mostly in high-activity nuclear waste. Plutonium (see page 356) is the most commonly produced actinoid, created when uranium isotopes are bombarded in nuclear fission reactors.

The atomic bombs dropped on Hiroshima and Nagasaki in 1945 owed their tremendous explosive power to fission of the unstable actinides uranium and plutonium.

# Transactinides

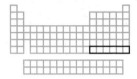

The transactinides are a series of metal elements all heavier than lawrencium (atomic number 103). They are sometimes referred to as transuranic elements, meaning elements heavier than uranium (92). All atoms of these superheavy elements are artificially produced by slamming together ions of different elements at high energies. They are so highly radioactive that they soon decay, often in fractions of a second (the most stable has a half-life of 28 hours). Needless to say, experiments are not feasible and most of their chemical properties are inferred theoretically. IUPAC – the International Union of Pure and Applied Chemistry – insists that an officially endorsed element must endure for at least $10^{-14}$ seconds. If it decays before this time, then it is considered never to have existed, since it had no time to form an electron cloud. Hence, there are claims pending for transactinides up to 118, but each must be painstakingly ratified. Glenn T. Seaborg (see page 114), who discovered both actinoid and transactinide series, also proposed a 'superactinide' series with atomic numbers between 121 and 155.

Target chamber in the Heavy Ion Research Centre (GSI) at Darmstadt, Germany.

# Atomic size

Finding the size of an atom is not as simple as putting it under a powerful microscope and measuring it. For a start, atoms exist in compounds. They shrink or swell depending on how they interact with other atoms – whether they are covalently bonded, exist as an ion or form part of a metallic lattice. Generally, half the bond length is used as a measure of atomic radius, but since covalently bonded atoms overlap their electron clouds, 'covalent radius' comes up smaller than 'metallic radius', and there is no way of directly comparing the relative sizes of non-metal and metal elements. Despite these technical difficulties, two things are clear: the size of atoms increases down a group and decreases along a period. The first trend is caused by the additional electron shells of heavier elements, but electron shielding, in which electrons in inner shells 'block' the electrostatic attraction outer electrons feel toward the nucleus, also plays a part. Along periods, electrons are added to the same valence shell. The shielding effect is negligible, so the extra proton in the nucleus draws the outer shell inwards.

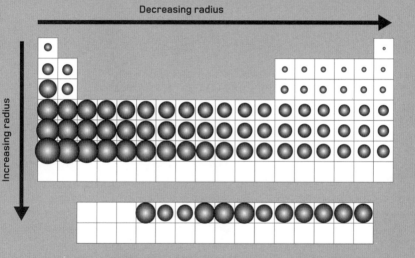

# Melting and boiling points

Melting points and boiling points are the temperatures at which substances change state – from solid to liquid, and liquid to gas respectively (see page 22). They can both be regarded as a measure of the energy of bonds holding a substance together, which must be overcome in order for a phase change to take place. For elements whose atoms form ionic and metallic bonds, melting and boiling points are high: their solids have strong electrostatic forces acting in all directions, and the attractive forces between individual atoms in liquids also tend to be strong. Covalently bonded substances with weak intermolecular forces, by contrast, need little encouragement to turn into liquids. Helium, a diatomic gas with negligible attraction between molecules, has the lowest melting point of all the elements, close to absolute zero. But beware of applying this rule universally – the element with the highest melting point of all, carbon in its diamond form, is also covalently bonded. Breaking diamond's strongly bonded crystal lattice requires a temperature of 3,642°C (6,588°F), at which point it sublimates, turning directly from solid to gas.

# Electronegativity

Electronegativity describes an element's affinity for electrons – its 'graspiness', or how easily it acquires electrons and binds with them. This is often described in terms of an 'ability' or 'willingness': although of course atoms have no agency, they are routinely thought to 'want' to achieve a full electron shell. In reality, of course, this is no more than energetics driving events: if a particular electron configuration is more stable, then a system will naturally 'settle' into it.

Because the metal elements on the left side of the periodic table have valence shells that are less than half full, it takes more energy for them to gain a full shell than it does to lose electrons and achieve stability that way. Therefore, these elements tend to lose electrons while those on the right side gain electrons. From left to right across periods, electronegativity increases, while it decreases down groups. The most electronegative element, therefore, is fluorine near the top-right corner. With full valence shells, the noble gases have no electronegativity values.

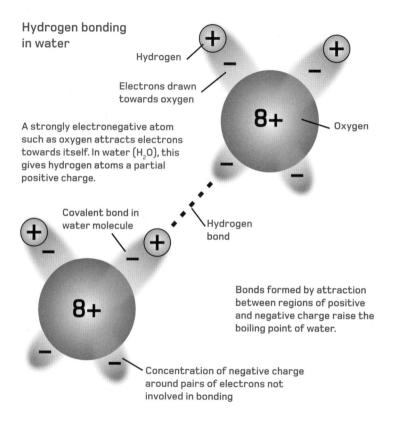

Hydrogen bonding
in water

Hydrogen

Electrons drawn
towards oxygen

A strongly electronegative atom
such as oxygen attracts electrons
towards itself. In water ($H_2O$), this
gives hydrogen atoms a partial
positive charge.

8+

Oxygen

Covalent bond in
water molecule

Hydrogen
bond

8+

Bonds formed by attraction
between regions of positive
and negative charge raise the
boiling point of water.

Concentration of negative charge
around pairs of electrons not
involved in bonding

# Ionization energy

Ionization energy is the energy required to remove a single electron from an isolated atom in its gaseous state – the energy needed to created a cation with a positive charge of +1. Unlike electronegativity (see page 160), which cannot be directly measured, the energy of first ionization is relatively simple to determine by experiment. The two concepts are different, but linked: the lower the ionization energy, the more easily an atom will form a cation; the higher the electronegativity, the more easily it will form an anion.

On the periodic table, ionization energy increases with atomic number along a period, as a result of the increasing valence shell stability. Down each group it decreases, largely because electrons in inner shells 'shield' valence electrons, reducing their attraction to the protons in the nucleus. These trends mirror electronegativity, but while noble metals have no value for electron affinity, they have very large ionization energies. This spike is due to the stability of a full valence shell.

## Ionization energies by atomic number

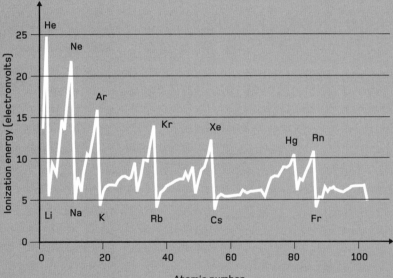

# Metallic character

$M$etals are elements whose atoms form metallic bonds, giving the bulk material the expected attributes of a metal, such as flexibility, workability and good conduction of heat and electricity. Despite the impression given by the periodic table, the line between metal and non-metal is not that clear; although the elements appear to be neatly sorted with metals facing non-metals like armies ranged against each other, the reality is more a grading of metal into non-metal.

From left to right along any period, metallic character tails off, but there is no sharp divide. The metals to the right of the d-block – sometimes called 'poor metals' – have only weak metallic character, while metalloids (see page 44) indulge in a variety of chemical bonding. Down a group, metallic nature increases, a pattern predicted by trends in ionization energy and electronegativity. As ionization energy decreases, so the tendency to form a positively charged cation increases: this explains the change from non-metal to metal in groups 13–15.

Ductility — the ability to draw out
a substance into an elongated
wire — is a key metallic property.

# Ferromagnetism

Iron, nickel and cobalt have an unusual property – they form powerful permanent magnets, especially when partnered with lanthanoid elements. The magnetic force has its origins in a fundamental property of electrons called spin, a form of angular momentum (rotation on an axis) that couples with their electric charge to create a small 'magnetic moment'. In order to share an electron orbital, electrons must have opposing spins, so electron pairs tend to cancel out each other. The unpaired electrons in ferromagnetic atoms, however, act like tiny bar magnets. They form 'domains' with aligned magnetic fields, but normally these domains are randomly oriented, so the material as a whole remains unmagnetized. Applying an external magnetic field causes the domains to snap into line, permanently magnetizing the material and amplifying the driving field. Heating a magnetic material above its so-called 'Curie temperature' mixes up the domains and resets its magnetism. Ferromagnetic elements are immensely important in technology – they are used in electricity generators, motors and magnetic storage technologies.

In bulk material, randomly orientated domains tend to cancel, leaving the substance unmagnetized.

When a magnetic field is applied to a ferromagnet, the domains line up permanently and the material is magnetized.

# Alternative
# periodic tables

**D**espite its status as an icon of science, the periodic table periodically comes in for criticism. Chief among the complaints is the charge that it has one job to do and doesn't do it very well: any periodic table should display graphically the core concept that the chemical elements show patterns that repeat in a particular sequence. However, the huge gaps at the top of the table fail to demonstrate this, and the sharp 'carriage returns' at the ends of the periods imply a discontinuity of elements.

Charles Janet's 'Left Step' table of 1928 was the first serious alternative periodic table. It orders elements in sequence of orbital filling, with some interesting effects. For example, helium becomes a group 2, rather than group 18, element. Theodor Benfey's 1964 spiral table (opposite) – often called the 'periodic snail' – arranges the elements on a winding unbroken band, allowing a greater appreciation of periodicity. Fernando Dufour's 3-D 'ElemenTree' of 1979 reveals a so-called 'secondary periodicity' of connections between elements in three dimensions.

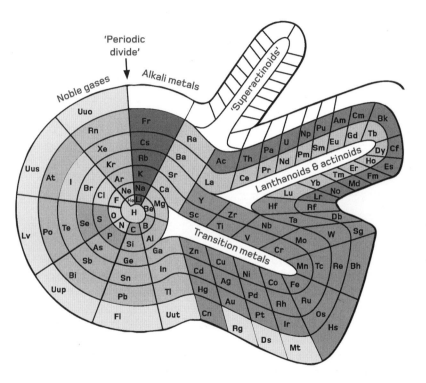

# Hydrogen

Hydrogen is the primordial element – the universe's original element and by far the most abundant. In 1815, English chemist William Prout hypothesized that all elements were made from whole-number combinations of hydrogen atoms (see page 58). His idea missed the beautiful complexity of the atomic nucleus, but its beguiling simplicity reveals a deep truth: through the process of nuclear fusion in stars, hydrogen begets all other elements (see page 90). Three-quarters of all ordinary matter in the universe – a full 90 per cent of all atoms – is hydrogen.

This light, colourless gas combines easily with almost every other element. Its single-proton nucleus easily loses its lone electron, dissociating from compounds in solution to make acids. This also makes hydrogen highly flammable, and it may ultimately save us from our dependence on fossil fuels – either by burning it directly, or by using it in fuel cells to generate electricity. Far too light to persist naturally on Earth as a gas, hydrogen is nevertheless abundant in water.

C ◉ ◯ H
Atomic radius: 53pm

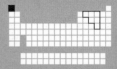

Group 1, Period 1
Non-metal

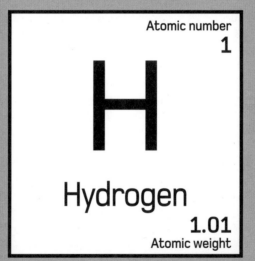

Atomic number
1

# H

## Hydrogen

1.01
Atomic weight

Melting point: −259.1°C (−434.5°F)
Boiling point: −252.9°C (−423.2°F)
Density: 0.00008988 g/cm³

Common isotopes: H–1, H–2
State (at STP): Gas
Colour: Colourless

# Helium

Helium is, quite literally, an otherworldly element. The signature of this strange gas was first discovered in the Sun in 1868, when Pierre Jules Janssen and Norman Lockyer independently analyzed the spectrum of sunlight (see page 104). Helium avoided detection on Earth until 1895, thanks to its scarcity and unreactivity as a noble gas (see page 150). This light element stubbornly refuses to form stable compounds with any other element, but can be collected alongside natural gas from rocks where it is trapped and prevented from escaping.

There is no lack of helium in the universe – it is the second-most abundant element, produced by hydrogen fusion reactions in the cores of stars. Best known for its use in party balloons, it is also a coolant *par excellence*: with the lowest melting point of all elements, liquid helium keeps superconducting magnets in MRI machines and particle accelerators chilled. Close to absolute zero, helium becomes a superfluid, with unusual properties such as the ability to flow upwards and even through solid objects.

C  He
Atomic radius: 31pm

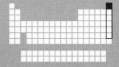

Group 18, Period 1
Noble gas

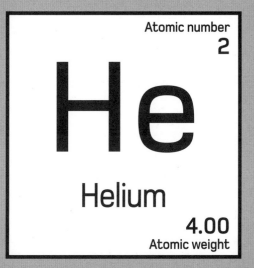

Atomic number
2

# He

Helium

4.00
Atomic weight

Melting point: –272.2°C (–457.9°F)
Boiling point: –268.9°C (–452°F)
Density: 0.0001785 g/cm³

Common isotopes: He-4, He-3
State (at STP): Gas
Colour: Colourless

# Lithium

The third of the elements created in the Big Bang (see page 88), lithium is the lightest metal and the least dense solid element: like several other group 1 elements, it floats on water. With just one electron in its outer shell, lithium is very reactive (although less so than its alkali metal brethren) and as a result is never found uncombined on Earth. Although comparatively rare, reserves are nevertheless widely distributed around the world.

Lithium has a variety of commercial uses. Its low density means that it combines with other metals to form lightweight alloys. Aircraft and rocket frames made from aluminium-lithium alloys help keep weight down, but still provide rigidity. The metal is also used as the anode in lithium batteries and in lithium ion rechargable batteries, which produce a greater voltage per unit volume than most other batteries. Despite mild toxicity, lithium chloride has a calming effect on the brain, and is prescribed to level out the troubling extremes of bipolar disorder.

C 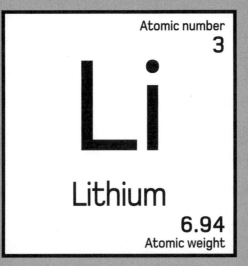 Li

Atomic radius: 167pm

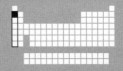

Group 1, Period 2
Alkali metal

Atomic number

3

# Li

Lithium

6.94
Atomic weight

Melting point: 180.5°C (357°F)
Boiling point: 1287°C (2349°F)
Density: 0.534 g/cm³

Common isotopes: Li-7, Li-6
State (at STP): Solid
Colour: Silver

# Beryllium

Beryllium is the element the universe forgot. Not produced in the grand explosion of matter that was the Big Bang, and skipped over by most of the fusion processes that take place in the cores of stars, element number 4 is rare and sparsely distributed. Despite this, it forms one of Earth's most beautiful gems – the beryllium aluminium silicate mineral, beryl.

Living organisms are generally reliant on the lighter elements, but thanks to its scarcity, beryllium is not found in the systems of living things – in fact, it is toxic even in the smallest quantities. Unlike its soft neighbours, beryllium is a hard metal and impervious to heat – yet while it is opaque to visible light, beryllium is transparent to X-rays. With only 500 tonnes extracted per year, it is used in situations where cost is no issue, such as astronautics and research. As well as superlight alloys, copper-beryllium alloy makes spark-proof tools for use in highly volatile atmospheres, such as oil wells.

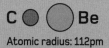 C ◯ ◯ Be

Atomic radius: 112pm

Group 2, Period 2
Alkaline earth metal

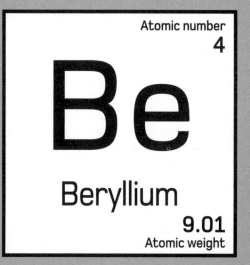

Atomic number

4

# Be

Beryllium

9.01

Atomic weight

Melting point: 1287°C (2349°F)
Boiling point: 2469°C (4476°F)
Density: 1.85 g/cm³

Common isotopes: Be-9
State (at STP): Solid
Colour: Grey

# Boron

The only metalloid in a group of otherwise metallic elements, boron is a rather unrepresentive figurehead for group 13. Other members of the boron group lose their three valence electrons willingly to form positive cations in metallic-bonded compounds. Smaller boron atoms, however, hold their electrons more tightly: hence the element forms covalent bonds and is a poor conductor of electricity. On Earth, boron is created by the action of high-energy cosmic rays from space. Although rare, it is found in many minerals of which borax, formed when salty lakes evaporate, is the most famous.

Like carbon and sulfur, boron has several allotropes: 'gamma boron', discovered in 2009, is nearly as hard as diamond. Boron oxide can be used to make tough borosilicate glass, used in ovenproof dishes and fibre-optic cable, while silicon 'doped' with boron is an important semiconducting material. Boron also stabilizes complex RNA molecules that are the precursors of DNA, and may have been critical in the evolution of life on Earth.

C 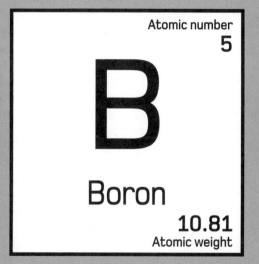 B

Atomic radius: 87pm

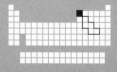

Group 13, Period 2
Metalloid

Atomic number
5

# B

## Boron

10.81
Atomic weight

Melting point: 2076°C (3769°F)
Boiling point: 3927°C (7101°F)
Density: 2.34 g/cm³

Common isotopes: B-11, B-10
State (at STP): Solid
Colour: Black

# Carbon

The fourth most abundant element in the universe, carbon only ranks at number 15 among the elements of Earth's crust. Yet all life ever found – from microscopic bacteria and protists to blue whales and giant redwood conifers – is built around carbon chemistry. There is clearly something special about this element, and it's not just a case of supply, but also versatility. Carbon can form single, double and triple covalent bonds, with itself and with other elements, placing it at the heart of a huge range of so-called 'organic' chemical compounds. Carbon's versatile bonding lends itself to an equally wide range of physical properties. Delocalized electrons within the carbon allotrope graphite give it metal-like properties such as electrical conductivity, while diamond, with the highest melting point of any known element, is decidedly non-metallic. Carbon bonds in linear chains, branching polymers, circular rings, flat sheets, tetrahedra and even spherical $C_{60}$ 'buckyballs'. Nanotubes of 'graphene' – an artifical carbon mesh just one atom thick – could make the strongest materials ever, with the potential to revolutionize engineering.

C 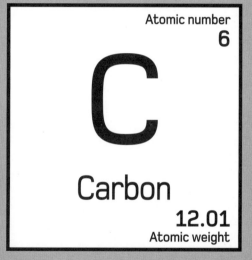 C
Atomic radius: 67pm

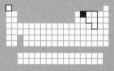

Group 14, Period 2
Non-metal

Atomic number
6

# C

## Carbon

12.01
Atomic weight

Melting point: 3527°C (6381°F)
Boiling point: 4027°C (7281°F)
Density: 2.267 g/cm³

Common isotopes: C-12, C-13
State (at STP): Solid
Colour: Black

# Nitrogen

Nitrogen is a colourless, odourless diatomic gas that makes up 78 per cent of Earth's air. We breathe it in all the time, but, since we have evolved to breathe a nitrogen-rich mix, it has no effect on us. Haemoglobin in blood binds to oxygen and draws it into the bloodstream but plenty of nitrogen gets in too: when divers surface too quickly, dissolved nitrogen forms bubbles in the blood. Forced into tissues, they cause an excruciating, potentially fatal, pain called 'the bends'.

Nitrogen itself is almost entirely inert, and is used where an atmosphere containing reactive oxygen would be dangerous or undesirable – for instance in welding, where sparks might cause explosions, or in food-packing plants. With its low melting point and plentiful supply, nitrogen is also used as a liquid coolant. It is an essential ingredient of DNA, but due to its unreactivity, it is tricky to properly absorb into the body. Certain 'nitrogen-fixing' plants, such as peas, have a symbiotic relationship with bacteria on their roots that helps them lock in nitrogen from the soil.

C 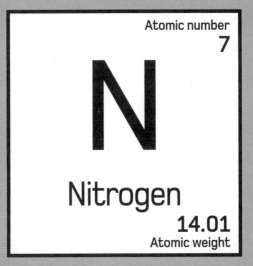 N
Atomic radius: 56pm

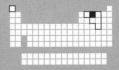

Group 15, Period 2
Non-metal

Atomic number
7

# N

## Nitrogen

14.01
Atomic weight

Melting point: -210°C (-346°F)
Boiling point: -195.8°C (-320.4°F)
Density: 0.0012506 g/cm³

Common isotopes: N-14, N-15
State (at STP): Gas
Colour: Colourless

# Oxygen

**D**iscovered in 1773 by Swedish chemist Carl Wilhelm Scheele (1742–86 – see page 82) oxygen is the most abundant element at the Earth's surface, and second overall (once iron in the core is taken into account). It is also our own main ingredient by weight: our bodies may run a carbon-based operating system, but this electronegative non-metal is the spark that keeps the machine ticking. In the universe, oxygen is the third element after hydrogen and helium – an abundance due to the 'doubly magic' stablility of the eight protons and eight neutrons in its nucleus.

Fiercely reactive, oxygen combines with almost any other element, snatching electrons from atoms to form oxides. Most oxygen on the planet is not in the air, but locked in solid oxides in Earth's crust. As dioxygen ($O_2$) gas, it gives the atmosphere an oxidizing effect, causing metals to tarnish and making combustion reactions possible (and hard to put out once they start). A thin layer of trioxide (ozone) gas, some 15–30 kilometres (9–18 miles) up, absorbs harmful high-energy ultraviolet rays from the Sun.

C 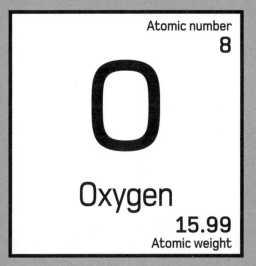 O
Atomic radius: 48pm

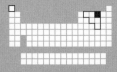

Group 16, Period 2
Non-metal

Atomic number
8

O

Oxygen

15.99
Atomic weight

Melting point: −218.8°C (−361.8°F)
Boiling point: −183°C (−297°F)
Density: 0.001429 g/cm³

Common isotopes: O-16, O-18, O-17
State (at STP): Gas
Colour: Colourless

# Fluorine

Heading up the halogens at the top of group 17 is fluorine. The most electronegative element on the periodic table, this pale yellow gas is a thoroughly nasty substance that will rip through almost any container holding it. Thanks to its relentless reactivity, fluorine has few uses as an isolated element, though it forms strong bonds with other elements and extremely stable compounds.

Polytetrafluoroethylene (PFTE), aka Teflon, excels at inactivity. It creates a water-repellant hydrophobic surface on fabrics, corrosion-resistant coatings for building materials and non-stick finishes on pots and pans. Chlorofluorocarbons (CFCs), meanwhile, were once used as coolants and aerosol propellants, but are now limited under the Montreal Protocol, due to the damage they cause to the ozone layer. The primary source of fluorine, the mineral fluorite, is used as flux for smelting metals – it lowers the melting temperature and viscosity of slag, allowing impurities to run off more easily.

C 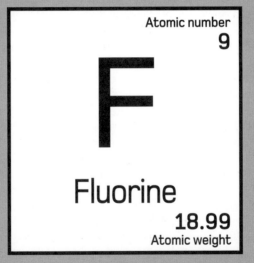 ⚫ ○ F
Atomic radius: 42pm

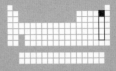

Group 17, Period 2
Halogen

Atomic number
9

# F

## Fluorine

18.99
Atomic weight

Melting point: -219.6°C (-363.3°F)
Boiling point: -188.1°C (-307°F)
Density: 0.001696 g/cm³

Common isotopes: F-19
State (at STP): Gas
Colour: Colourless

# Neon

No element is more closely identified with a single use than neon, a noble gas that has permanently changed our mental image of a cityscape. Neon produces an intense orange-red light when a high voltage is applied across low pressure gas – other colours can be produced by phosphorescent coatings on the glass tube. Flashing neon signs with their buzzing tubes and garish colours are now an essential part of any city. The technology was developed originally in France around 1910 but came into its own in the USA, particularly in New York City, where it was marketed as 'liquid fire'.

A colourless and odourless monatomic gas, neon has the most restricted liquid phase of all elements, between -248.45 and -245.95°C (-415.46 and -410.88°F). Although it is the fifth most common element in the universe, it is rare on Earth, due to its inability to form compounds (it is the least reactive element on the periodic table). Unable to form bonds, this lighter-than-air gas drifts away and escapes the atmosphere into space.

C ⬤ ○ Ne
Atomic radius: 38pm

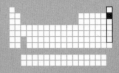

Group 18, Period 2
Noble gas

Atomic number
10

# Ne

Neon

20.18
Atomic weight

Melting point: -248.6°C (-415.5°F)
Boiling point: -246.1°C (-411°F)
Density: 0.0008999 g/cm³

Common isotopes: Ne-20, Ne-22, Ne-21
State (at STP): Gas
Colour: Colourless

# Sodium

L ike so many of the group 1 and 2 metals, sodium has a split
personality. On one side is the pure, isolated metal, volatile
and dangerous, while on the other, are the numerous benign and
beneficial minerals and salts formed by this reactive element.
Pure sodium is a soft, silvery metal that tarnishes immediately
when exposed to air and reacts explosively with water, producing
hydrogen. Unsurprisingly, the pure metal does not exist in nature,
so it's perhaps surprising that some nuclear reactors use
molten sodium to remove heat from the reactor core.

Sodium ions are also critical for the correct functioning of the
human body. Blood plasma and extracellular fluid bathe cells in a
sodium-rich salt solution permitting transport in and out of the
cell. Sodium and potassium ions also mediate nerve signals (see
page 206). Most sodium comes from dietary salt (sodium chloride).
On average, humans require 500 mg of sodium per day, but some
people routinely take much more, potentially to the detriment of
their health. Salt is also an important industrial chemical.

C Na
Atomic radius: 190pm

Group 1, Period 3
Alkali metal

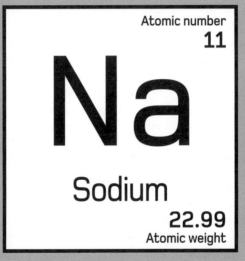
Atomic number
11

Na

Sodium

22.99
Atomic weight

Melting point: 97.7°C (207.9°F)
Boiling point: 883°C (1621°F)
Density: 0.971 g/cm³

Common isotopes: Na-23
State (at STP): Solid
Colour: Silver

# Magnesium

One common recollection from school chemistry classes is the effect of burning magnesium ribbon. The blinding white light that flares up when the metal ignites is nothing if not memorable. It is also nearly impossible to extinguish, since burning magnesium reacts exothermically not only with oxygen, but also with nitrogen and water. This unusual property was used to devastating effect when magnesium was incorporated into the casings of incendiary bombs during the Second World War. In bulk, however, magnesium is quite difficult to ignite, and when partnered with aluminium, it makes an easy-welding, light alloy (although there have been several cases of fatalities caused by alloy racing cars literally crashing and burning).

Magnesium is abundant in the Earth's crust and mantle rocks, and an essential element in biology. Its integral roles in plant chlorophyl, the genetic DNA and RNA molecules, the energy-giving ATP compound and many enzymes give it a claim to be the most important element for life on Earth.

C ⬤ ◯ Mg

Atomic radius: 45pm

Group 2, Period 3
Alkaline earth metal

Atomic number
12

# Mg

## Magnesium

24.30
Atomic weight

Melting point: 650°C (1202°F)
Boiling point: 1090°C (1994°F)
Density: 1.738 g/cm³

Common isotopes: Mg–24, Mg–26, Mg–25
State (at STP): Solid
Colour: Silver

# Aluminium

Aluminium is a light metal, and the most abundant in the Earth's crust. However, it is tightly locked up in minerals and is energy-intensive to extract. Indeed, it was once considered a precious metal alongside gold and silver (Emperor Napoleon served his guests with aluminium tableware). However, thanks to the Hall–Héroult process (see page 110), it is now used in everything from windows to automobile bodies and cooking foil. An estimated 50 aluminium cans are manufactured every second.

As well as exceptional lightness, aluminium is known for its corrosion resistance. Unlike many metals (such as iron, which flakes when oxidized, exposing fresh surfaces to attack), aluminium forms an impenetrable oxide coating on exposure to air. Off the main block of transition metals, this 'poor' metal is prone to ripping under stress, and benefits from alloying. Although not generally thought of as reactive, aluminium in powdered form burns fiercely. It is used in solid rocket propellant and in pyrotechnic 'flash powder'.

C 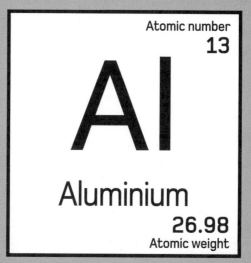 Al
Atomic radius: 118pm

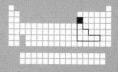

Group 13, Period 3
Post-transition metal

Atomic number
13

Al

Aluminium
26.98
Atomic weight

Melting point: 660.3°C (1220.6°F)
Boiling point: 2519°C (4566°F)
Density: 2.698 g/cm³

Common isotopes: Al-27
State (at STP): Solid
Colour: Silver

# Silicon

A metalloid like so many of the carbon group elements, silicon has an unusual mix of properties. As the most abundant element in the Earth's crust, silicon is a ubiquitous rock-former. As silica, bonded with two oxygen atoms, it forms a full 90 per cent of all minerals and is the basis of glass.

Like carbon, silicon has four valence electrons available to form bonds, and much has been made of its potential as a foundation for alien life. But given its abundance on our planet, why are we not silicon-based lifeforms ourselves? The explanation may lie in the fact that silicon's outer electron shell lies farther from the nucleus, and therefore tends to form weaker bonds. A few groups of organisms – notably sponges and microscopic radiolarians – use silicon to build their bodies, but most have opted for calcium phosphate instead. Perhaps silicon is more likely to create artificial intelligence than organic life. Ultra-pure crystals can be etched with electronic circuits, making 'chips' that hold billions of semiconducting components in an area the size of a fingernail.

C 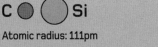 Si

Atomic radius: 111pm

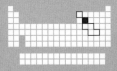

Group 14, Period 3
Metalloid

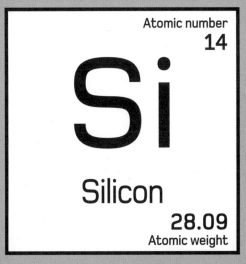

Atomic number
14

## Si

Silicon

28.09
Atomic weight

Melting point: 1414°C (2577°F)
Boiling point: 3265°C (5909°F)
Density: 2.3296 g/cm³

Common isotopes: Si-28, Si-29, Si-30
State (at STP): Solid
Colour: Grey

# Phosphorus

Discovered in 1669 by Hennig Brand (see page 56), this non-metal was the first element to be isolated by chemical means. Brand rather stumbled across the most volatile of the element's allotropes – white phosphorus, which burns spontaneously on contact with air – hence why it was soon named after the Greek for 'light bringer'.

The earliest matches – called 'lucifers' – had white phosphorus tips, but factory workers began to die from exposure to its toxic effects. Today's strike-anywhere matches use a less feisty allotrope – red phosphorus. Lethal as pure phosphorus may be, the phosphate ion ($PO_4^{3-}$) is ubiquitous in the human body and essential to life. As an ionic compound with calcium, phosphate forms the hard mineral content of bones and teeth, locking up most of the body's 750 grams (26.5 oz) of phosphorus. However sugar phosphate molecules also form the 'side rails' of the ladder-like DNA molecule, and phosphate transfer between molecules in cells powers the body's main energy system.

C 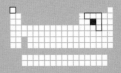 P

Atomic radius: 98pm

Group 15, Period 3
Non-metal

Atomic number
15

P

Phosphorus
30.97
Atomic weight

Melting point: 44.2°C (111.5°F)
Boiling point: 277°C (531°F)
Density: 1.82 g/cm³

Common isotopes: P-31
State (at STP): Solid
Colour: Colourless

# Sulfur

With more than 30 solid forms, sulfur is the periodic table's undisputed allotrope supremo. In its native state, it encrusts the mouths of volcanic fumaroles with sickly yellow crystals. The acrid smell of sulfurous gases hissing out of these vents is violently unpleasant, so it is no surprise that sulfur has a long-standing association with the underworld, and was associated Biblically with hellish 'brimstone'.

Despite these connotations, however, sulfur is relatively benign. It is an essential element for life, used by animals during protein synthesis, and concentrated in bird eggs – the pungent 'rotting egg' smell is hydrogen disulfide. 'Vulcanizing' rubber, by adding sulfur to it, makes it more pliable and durable – a process that allowed the development of rubber tyres and inner tubes. Sulfur dioxide, meanwhile, is the precursor to sulfuric acid, the world's most important industrial chemical. However, it also causes acid rain and, as an unwanted emission from burning fossil fuels, is now a major industrial pollutant.

C 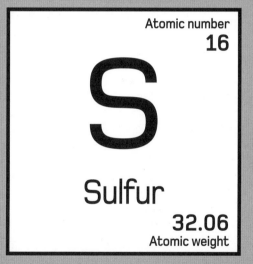 S
Atomic radius: 88pm

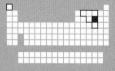

Group 16, Period 3
Non-metal

Atomic number
16

# S

## Sulfur

32.06
Atomic weight

Melting point: 115.2°C (239.4°F)
Boiling point: 445°C (833°F)
Density: 2.067 g/cm³

Common isotopes: S-32, S-34, S-33
State (at STP): Solid
Colour: Yellow

# Chlorine

Quick and deadly, element number 17 can never free itself from its associations with warfare. Allegedly touted by German chemist Fritz Haber as a 'higher form of killing', chlorine gas was used on First World War battlefields: the very same properties that make it a ferocious cleaning agent and disinfectant in bleaches make it a terrible weapon. Taken into the lungs, this extremely reactive halogen wreaks havoc on the delicate tissues, leaving victims drowning in their own blood. More benignly, chlorine can be used to kill pathogens in water, purifying it for drinking or use in swimming pools.

Under normal conditions, chlorine is a diatomic yellow-green gas. Single atoms are highly electronegative and swiftly form ionic salts, 'stealing' an electron from other atoms to fill their outer shell. The most important salt is sodium chloride (table salt), which plays a vital role in the body. Chlorine is also present in a wide range of important industrial compounds, including hydrochloric acid, plastics such as PVC and the insecticide DDT.

C 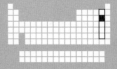 Cl
Atomic radius: 79pm

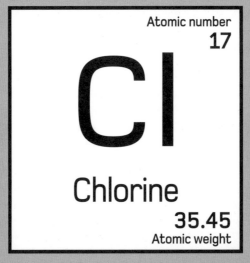

Atomic number
17

Cl

Chlorine

35.45
Atomic weight

Melting point: -101.6°C (-150.8°F)
Boiling point: -34°C (-29°F)
Density: 0.003214 g/cm³

Common isotopes: Cl-35, Cl-37
State (at STP): Gas
Colour: Yellow

# Argon

Although it's often overlooked and has a name that means 'lazy', argon makes up just under 1 per cent of the Earth's atmosphere. That equates to some 50 trillion tonnes of the gas floating in the air, and makes it the third most abundant atmospheric element after nitrogen and oxygen.

As a typical noble gas, argon is almost entirely unreactive, and therefore its normal state is the monatomic, elemental form. Argon's unreactivity does not render it useless, however – it is often used to create a safe inert atmosphere for substances that are prone to explode or react dangerously. Along with oxygen, it is also bubbled through molten steel – the argon stirs the metal while the oxygen removes carbon in the form of carbon dioxide. Argon is used to exclude air and avoid oxidation when arc-welding aluminium and growing super-pure silicon crystals. In 2000, chemists at the University of Helsinki finally persuaded it to take part in a reaction; however, the resulting argon fluorohydride was unstable at temperatures above $-246$ °C ($-411$ °F).

C 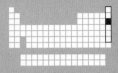 ⬤ ◯ Ar
Atomic radius: 71pm

Group 18, Period 3
Noble gas

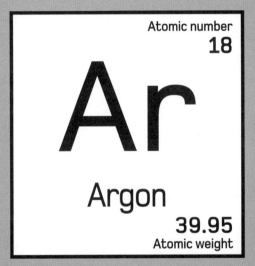

Atomic number
18

Ar

Argon

39.95
Atomic weight

Melting point: -189.4°C (-308.8°F)
Boiling point: -185.9°C (-302.6°F)
Density: 0.0017837 g/cm³

Common isotopes: Ar-40, Ar-36, Ar-38
State (at STP): Gas
Colour: Colourless

# Potassium

The elements of group 1 get progressively more reactive down the group. Accordingly, potassium is a fiery and temperamental metal. Soft, like the other alkali metals, and quick to tarnish, potassium must be stored under oil to stop it reacting with air. Chemically, it is the twin of sodium, but it is the more reactive of the pair: potassium's reaction with water is so intense that the hydrogen gas coming off the water explodes. The metal burns with a brilliant, distinctive lilac flame. It may be surprising to find such a volatile element in the body, but potassium is essential for transmitting nerve signals around the brain and body. The average 70-kilogram (155-lb) person contains around 140 grams (5 oz) of potassium, and most of it comes from fruit and vegetables. Bananas contain notable amounts, of which some is radioactive potassium-40 (the isotope's decay produces most of the argon in the Earth's atmosphere). This has led to the whimsical designation of the 'banana equivalent dose' — the dose of ionizing radiation received from eating one banana.

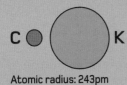

C ◯ K

Atomic radius: 243pm

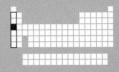

Group 1, Period 4
Alkali metal

Atomic number
19

# K

Potassium

39.09
Atomic weight

Melting point: 63.4°C (146.1°F)
Boiling point: 759°C (1398°F)
Density: 0.862 g/cm³

Common isotopes: K–39, K–41, K–40
State (at STP): Solid
Colour: Silver

# Calcium

A soft, shiny and reactive metal, silver-grey calcium is rarely encountered in its pure form. Instead, this reactive element is a great builder, readily forming ionic salts. Element 20 is the fifth most common in the Earth's crust, where it is mostly found as calcium carbonate in limestone and chalk rocks. We carry more than a kilogram (2.2 lb) of it around as calcium phosphate in our bones, while quicklime, created by heating calcium carbonate, is the key ingredient in the 5.2 billion tonnes of concrete cement produced each year.

Calcium compounds dissolve easily in water, especially when it is slightly acidic. Although limestone is a hard-wearing rock, it is not weatherproof and is eaten away by rain and groundwater. The calcium cations released into the water make it 'hard' – prone to leaving limescale deposits and resistant to making suds with soap. Sea creatures build their shells out of dissolved calcium carbonate, extracted from the water. These tough structures persist long after the animal dies, and can become fossils.

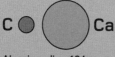

C ◯ ⬤ Ca

Atomic radius: 194pm

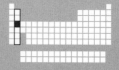

Group 2, Period 4
Alkaline earth metal

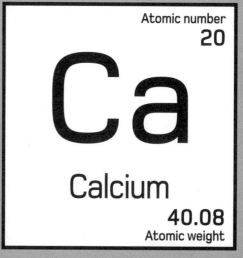

Atomic number
20

# Ca

Calcium

40.08
Atomic weight

Melting point: 842°C (1548°F)
Boiling point: 1484°C (2703°F)
Density: 1.54 g/cm³

Common isotopes: Ca-40, Ca-44, Ca-42
State (at STP): Solid
Colour: Silver

# Scandium

When Dmitri Mendeleev prepared his periodic table of 1869 (see page 68), he took the brave step of predicting the positions (and atomic masses) of new undiscovered elements. *Eka*-boron, with a predicted atomic mass of around 44, was the lightest of these candidates. The identification of an element with the right properties by spectral analysis in 1879, hard on the heels of the 1875 confirmation of *eka*-aluminium (gallium – see page 230), showed the power of Mendeleev's proposition.

Scandium was first identified in gadolinite and euxenite minerals from Scandinavia, and takes its name from that region. It is about as abundant as lead, but harder to extract since it doesn't accumulate in appreciable deposits. As one of only two group 3 metals – the other being the equally Nordic yttrium – it is often counted as a rare earth metal, along with the lanthanoids. Scandium is a relatively soft metal, but when added to aluminium it has a major strengthening effect, making strong and light alloys used in bike frames and aerospace parts.

Atomic radius: 184pm

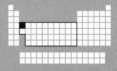

Group 3, Period 4
Transition metal

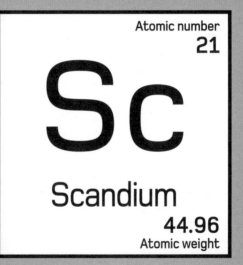

Atomic number
21

# Sc

Scandium

44.96
Atomic weight

Melting point: 1541°C (2806°F)
Boiling point: 2836°C (5137°F)
Density: 2.989 g/cm³

Common isotopes: Sc-45
State (at STP): Solid
Colour: Silver

# Titanium

Colossally strong, fantastically tough and bearing the mythical name of the Titans (rivals to the gods of Mount Olympus in Greek mythology), titanium is a true superhero element. Few elements have penetrated popular culture in quite the same way as this white, shiny transition metal – it has become synonymous with qualities of resilience, even featuring in pop songs.

With the highest strength-to-weight ratio of all the elements, titanium alloys are used in jet engines, spacecraft and lightweight sports equipment (although the element enthusiast Theodore Gray recommends taking an angle grinder to your 'titanium' golf clubs to test for the telltale bright white sparks). Like tantalum (see page 314), titanium won't react with anything inside the body and so it can be used for replacement hips, dental implants and pins to hold broken bones together. Titanium is not rare, but the costs of refining keep it reassuringly expensive. Most is found in the form of titanium dioxide – the brilliant white solid used in paints and to whiten paper.

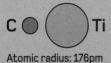

C ◯ ⬤ Ti

Atomic radius: 176pm

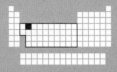

Group 4, Period 4
Transition metal

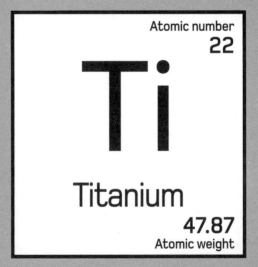

Atomic number
22

Ti

Titanium

47.87
Atomic weight

Melting point: 1668°C (3034°F)
Boiling point: 3287°C (5949°F)
Density: 4.54 g/cm³

Common isotopes: Ti-48, Ti-46, Ti-47
State (at STP): Solid
Colour: Silver

# Vanadium

Another element with a Scandinavian connection, vanadium is named for Vanadis – one of the Norse goddess Freyja's nine additional names. Fittingly, it is renowned for its strength and beauty. Element 23 is a hard, steel-blue metal that is unreactive and resistant to corrosion. Just a small addition of vanadium to steel gives a great increase in tensile strength (resistance to stretching forces) and hardness. Chrome-vanadium steel is used for making tools, although vanadium is actually just one part of a multi-component system that also includes manganese, phosphorus, sulfur, silicon and chromium.

A classic transition metal, vanadium has several 'oxidation states', or configurations of its outer electrons. This offers a wide choice of options for bonding, and the ability to form complex ions – a metal cation surrounded by loosely bonded neutral molecules – which typically produce brightly coloured compounds. Vanadium pentoxide is an important industrial catalyst used in the contact process to make sulfuric acid.

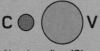
Atomic radius: 171pm

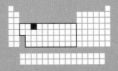
Group 5, Period 4
Transition metal

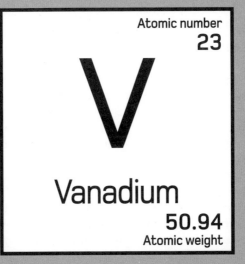

Atomic number
23

V

Vanadium

50.94
Atomic weight

Melting point: 1910°C (3470°F)
Boiling point: 3407°C (6165°F)
Density: 6.11 g/cm³

Common isotopes: V–51, V–50
State (at STP): Solid
Colour: Silver

# Chromium

Flashy and corrosion resistant, chromium was a post-war vision of the future. The technique of applying a thin layer of protective chromium metal to a steel surface had been developed in the 1920s, but it wasn't until the 1940s that it became widespread. The USA became a chrome-plated marvel-land of gleaming streamlined toasters and automobile tailfins. Meanwhile, another chromium technology was taking over. Stainless steel – made of steel mixed with 10 per cent by mass of chromium – was cheap to produce, resistant to rusting and much less likely to peel off. It is still the alloy of choice for surgical instruments, gleaming steel exteriors and cutlery.

Chromium was named in 1797 by Louis Vauquelin after the vividly coloured compounds he produced from its mineral ore. As an impurity, chromium puts the green in emeralds, and the metal was used to produce the 'chrome yellow' pigment used by 19th-century artists. Unusually, chromium (III) is a vital trace element in the body, but chromium (VI) is a toxic heavy metal.

C ○ ◯ Cr

Atomic radius: 166pm

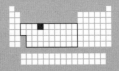

Group 6, Period 4
Transition metal

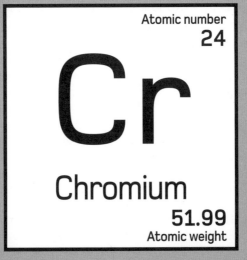

Atomic number
24

# Cr

Chromium

51.99
Atomic weight

Melting point: 1907°C (3465°F)
Boiling point: 2671°C (4840°F)
Density: 7.15 g/cm³

Common isotopes: Cr-52, Cr-53, Cr-50
State (at STP): Solid
Colour: Silver

# Manganese

Manganese is a hard and brittle metal, used principally in steel alloys. Despite its low profile, it is the third-most abundant transition element in the Earth's crust, after iron and titanium, and most modern steel is manganese steel. Small amounts of the metal added to steel increase its workability at high temperatures, while larger quantities (between 8 and 15 per cent) increase the tensile strength of the solid alloy: steel helmets worn by construction workers and soldiers both contain significant amounts of the metal. Manganese dioxide mixed with carbon powder, meanwhile, forms the positive terminal in widely used alkaline batteries.

As a trace element essential for the correct function of many enzymes in the body, manganese is required in amounts of 5 mg daily. Double that dose, however, and the element becomes a deadly neurotoxin, causing 'manganese madness'. With symptoms similar to Parkinson's disease, manganism also comes with a side-order of psychosis.

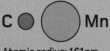

Atomic radius: 161pm

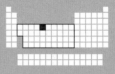

Group 7, Period 4
Transition metal

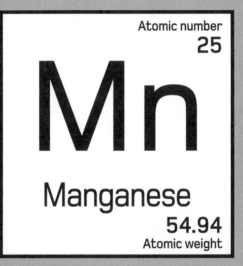

Atomic number
25

# Mn

Manganese
54.94
Atomic weight

Melting point: 1246°C (2275°F)
Boiling point: 2061°C (3742°F)
Density: 7.44 g/cm³

Common isotopes: Mn-55
State (at STP): Solid
Colour: Silver

# Iron

In the hearts of massive stars, fusion of nickel-56 is the final nuclear fusion reaction that is accompanied by a release of binding energy (see page 16). As a result, it is the last element to be synthesized before the star runs out of fuel and spews its contents across space in a violent supernova. Nickel-56 decays to iron-56, so iron sits at the top of the hill as the most stable element in the universe. This element likes to be at the heart of things: thanks to intense radiation from the young Sun that drove away lighter elements, iron is concentrated in the inner planets of the solar system. It is the most plentiful element on Earth, but is concentrated in the core (see page 92). In the haemoglobin molecule, iron delivers oxygen in the blood to every cell of the body. You require a daily dose of 20 mg to make new red blood cells and your body contains enough iron to make a sizeable (7.5-centimetre/3-in) nail. Mastery of iron and steel has driven developments in human technology for more than 2,000 years, and the element is also useful for its magnetic properties.

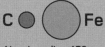

C Fe

Atomic radius: 156pm

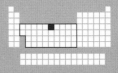

Group 8, Period 4
Transition metal

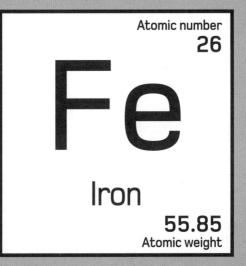

Atomic number
26

Fe

Iron

55.85
Atomic weight

Melting point: 1538°C (2800°F)
Boiling point: 2861°C (5182°F)
Density: 7.874 g/cm³

Common isotopes: Fe-56, Fe-54, Fe-57
State (at STP): Solid
Colour: Grey

# Cobalt

The medieval silver miners of Saxony imagined themselves plagued by *kobold* – subterranean sprites who shared their mines, causing rock falls and collapses. Their favourite trick, however, was to swap good ores for worthless drek. Cobalt took its name from these mischievous troublemakers, since its minerals held few precious metals.

Cobalt is one of three ferromagnetic elements on the periodic table, and the one with the highest Curie temperature (see page 166). It adds a deep blue colour to glass – a trick that was discovered in China more than 2,000 years ago – and is used in pottery glazes. Cobalt has one stable isotope, cobalt-59. The other, cobalt-60, is a source of gamma radiation and is mostly used for treating certain cancers. Co-59 is turned into Co-60 by bombarding it with 'slow' neutrons in a nuclear reactor. Leó Szilárd – designer of the nuclear fission chain reaction – noted that a 'dirty bomb' consisting of a nuclear explosive with a Co-59 jacket had the potential to poison land for decades.

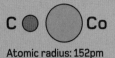

C ◯ ◯ Co

Atomic radius: 152pm

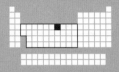

Group 9, Period 4
Transition metal

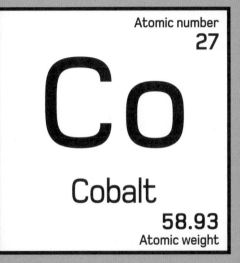

Atomic number
27

## Co

Cobalt

58.93
Atomic weight

Melting point: 1495°C (2723°F)
Boiling point: 2927°C (5301°F)
Density: 8.86 g/cm³

Common isotopes: Co-59
State (at STP): Solid
Colour: Grey

# Nickel

Like many d-block metals, nickel is a superior alloy-maker. It produces heat-resistant superalloys, used in aircraft turbines and rocket boosters. On Earth, however, most nickel is out of reach. Its four-billion-year parnership with iron in the planet's core leaves the crust depleted of the transition metal.

Nickel is corrosion resistant (it does tarnish in air, but very slowly) and can be polished to a beautiful sheen. Because it is cheaper to produce, it's now used as a super-shiny plating layer more widely than chromium. As the third of the periodic table's triumvirate of ferromagnetic elements (see page 166), it is used to make 'alnico' (an aluminium/nickel/cobalt alloy) electromagnets, with a strength between those of iron permanent magnets and rare-earth magnets. Although largely inert in bulk, the element is actually fairly reactive when its surface area is increased: powdered nickel catalyzes the hydrogenation reactions that produce margarine industrially, and foamed nickel is used as a low-density anode in certain fuel cells.

C 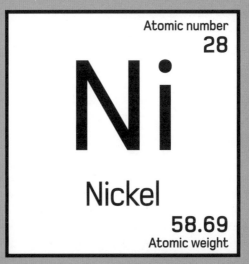 Ni

Atomic radius: 149pm

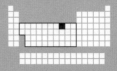

Group 10, Period 4
Transition metal

Atomic number
28

# Ni

Nickel

58.69
Atomic weight

Melting point: 1455°C (2651°F)
Boiling point: 2913°C (5275°F)
Density: 8.912 g/cm³

Common isotopes: Ni-58, Ni-60, Ni-62
State (at STP): Solid
Colour: Grey

# Copper

Copper straddles the ancient and modern worlds. Found both as a native element and an ore, this unusual-looking reddish-brown metal was one of the first to be extracted from its minerals (see page 94). Though too soft on its own for rugged uses, when added to tin in a 2:1 ratio it forms bronze – a much tougher, workable alloy that holds an edge. This discovery, around 2500 BCE, was a major technological innovation in human history and ushered in the Bronze Age.

Today, however, copper has come into its own as a conductor of electrical current. It forms the backbone of modern power grids and the conductive tracks on electronics circuit boards. Copper's value means that it is now the third most recycled metal after iron and aluminium – an estimated 80 per cent of copper produced is still in use. The antimicrobial properties of copper have also recently been rediscovered: copper push plates, rails and bed knobs are being retrofitted in hospitals to discourage the harbouring and transmission of superbugs.

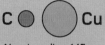

C ○ ● Cu

Atomic radius: 145pm

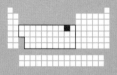

Group 11, Period 4
Transition metal

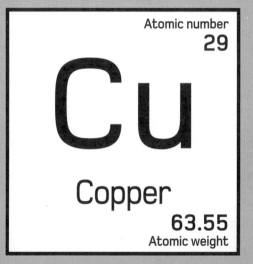

Atomic number
29

# Cu

Copper

63.55
Atomic weight

Melting point: 1084.6°C (1984.3°F)
Boiling point: 2562°C (4644°F)
Density: 8.96 g/cm³

Common isotopes: Cu-63, Cu-65
State (at STP): Solid
Colour: Copper

# Zinc

Zinc's finest hour came in 1800, when it became one of the two terminals in Alessandro Volta's 'voltaic pile' – the world's first chemical battery. It is still used as the anode in many batteries, but rather than taking a starring role, zinc is a utility player these days, in service to other metals. Glitzy brass is zinc alloyed to copper. Even the US cent is 97.5 per cent zinc with a copper cladding to keep up appearances.

Over half of the zinc produced goes to galvanize steel. In this process, steel gets a protective coating of zinc, either by hot dip or electroplating. Zinc is mildly reactive and combines with carbon dioxide in moist air to form zinc carbonate – a dull grey unreactive layer that seals the surface from further corrosion. Zinc's chemistry is less varied than many other transition metals, and many consider group 12 to be the first of the post-transition metals (see page 136). Its compounds are mostly formed by positive 2+ cations. White zinc oxide is ubiquitous in paint pigments and UV-blocking suncreams.

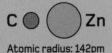
C ◯ ⬤ Zn

Atomic radius: 142pm

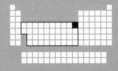

Group 12, Period 4
Transition metal

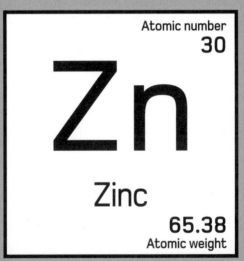

Atomic number
30

# Zn

Zinc

65.38
Atomic weight

Melting point: 419.7°C (787.5°F)
Boiling point: 907°C (1665°F)
Density: 7.134 g/cm³

Common isotopes: Zn-64, Zn-66, Zn-68
State (at STP): Solid
Colour: Slate grey

# Gallium

This soft, silvery metal has a famously low melting point; spoons made out of it will 'disappear' when used to stir a hot drink and it even melts in the hand. Most, however, is used to make semiconducting gallium arsenide and gallium nitride – used in ultra-high-speed logic chips and laser diodes.

Gallium was discovered in 1875 by a French chemist by the name of Paul-Émile Lecoq de Boisbaudran (1838–1912) who mounted a quest to identify new elements from their spectra. The new element was significant because it slotted into the space of Dmitri Mendeleev's predicted *eka*-aluminium (see page 68) and was a crucial piece of evidence for the periodic law of chemical elements (see page 62). It also ignited the greatest priority dispute since the Anglo-French brouhaha over the discovery of oxygen. Mendeleev felt that the new metal merely confirmed his discovery and the credit should be his. But the Frenchman may have had the last laugh – gallium refers to France (*Gallia* in Latin) but also obliquely to its discoverer's name.

C 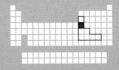 Ga

Atomic radius: 136pm

Group 13, Period 4
Post-transition metal

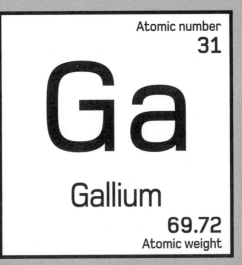

Atomic number
31

## Ga

Gallium

69.72
Atomic weight

Melting point: 29.8°C (85.6°F)
Boiling point: 2204°C (3999°F)
Density: 5.907 g/cm³

Common isotopes: Ga-69, Ga-71
State (at STP): Solid
Colour: Silver

# Germanium

G ermanium is a brittle metalloid, similar in appearance to silicon, and discovered in 1886 by the chemist Clemens Winkler (1838–1904). Named after Germany, its discovery strengthened Mendeleev's periodic law of chemical elements, fitting into the space where he had predicted *eka*-silicon.

Like all carbon group elements, germanium has four valence electrons. 'Doping' germanium crystals with pentavalent elements, such as arsenic, creates an 'n-type' donor semiconductor, with a surfeit of electrons; doping with trivalent aluminium or indium makes a 'p-type' acceptor semiconductor with 'holes'. Cheap solid-state germanium transistors fuelled a post-war boom in consumer electronics. Now largely superseded by silicon-based electronics, germanium semiconductors are still used in some wireless devices and guitarists swear by the vintage tones they produce. Germanium dioxide, meanwhile, is an important industrial chemical, used in fibre-optic cable, and also to catalyze the polymerization of PET plastics used for plastic bottles.

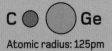

C ◯ ◯ Ge

Atomic radius: 125pm

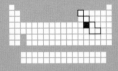

Group 14, Period 4
Metalloid

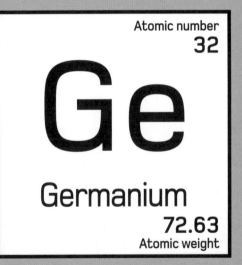

Atomic number
32

Ge

Germanium

72.63
Atomic weight

Melting point: 938.3°C (1720.9°F)
Boiling point: 2833°C (5131°F)
Density: 5.323 g/cm³

Common isotopes: Ge-74, Ge-72, Ge-70
State (at STP): Solid
Colour: Grey

# Arsenic

This toxic element sits directly beneath phosphorus on the periodic table. Arsenic's notoriety comes from its long history as 'inheritance powder' – arsenic trioxide, the poisoner's weapon of choice. Its reign of terror only came to an end in 1836 with the invention of the Marsh test, which detects minute quantities of arsenic in a cadaver.

Since phosphorus is so important in living systems (see page 198), it is somewhat surprising that arsenic is so deadly. In 2010, however, element 33 hit the headlines, with claims from NASA scientists that bacteria living in highly saline lakes substituted arsenic for phosphorus in their DNA. Since life operates a monoculture of carbon-based organisms, such a finding would have been equivalent to the discovery of a second genesis. The claim was debunked in 2012, when it was shown that although the GFAJ-1 bacterium was indeed highly tolerant of arsenic, it went to great lengths to procure what phosphorus it could from the environment: there was no mystery after all.

C ⬤ ◯ As

Atomic radius: 114pm

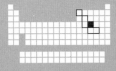

Group 15, Period 4
Metalloid

Atomic number
33

As

Arsenic

74.92
Atomic weight

Melting point: 817°C (1503°F)
Boiling point: 614°C (1137°F)
Density: 5.776 g/cm³

Common isotopes: As-75
State (at STP): Solid
Colour: Silver

# Selenium

Selenium is a metalloid element, discovered in 1817 by Jöns Jakob Berzelius and Johann Gahn (1745–1818). It betrayed its presence in a pyrite (sulfite) ore by smelling of horseradish – a trait common to all selenium compounds. Tellurium, a period lower in the oxygen group, is also smelly (see page 272), so Berzelius named selenium after the Greek Moon goddess, to link it with its earthy twin. Selenium is still produced from sulfide ores.

Like so many metalloid elements, selenium has several allotropes. Grey selenium is a shiny metal that exhibits the photoelectic effect, where light striking a surface can liberate electrons; brick-red selenium is an amorphous non-metal. It is one of those puzzling chemical elements that are essential to the correct functioning of the human body in tiny amounts, but toxic in barely greater quantities (over 0.4 mg). As a result, a normal diet easily provides sufficient selenium and deficiency is rare. Brazil nuts and peaches are rich in selenium, but the quantities in foodstuffs mostly depend on the selenium content of soils.

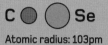

C ◯ ◯ Se

Atomic radius: 103pm

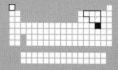

Group 16, Period 4
Non-metal

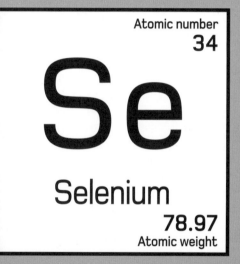

Atomic number
34

# Se

Selenium

78.97
Atomic weight

Melting point: 180°C (356°F)
Boiling point: 685°C (1265°F)
Density: 4.809 g/cm³

Common isotopes: Se-80, Se-78, Se-76
State (at STP): Solid
Colour: Grey

# Bromine

There are only two liquid elements on the periodic table at standard temperature and pressure— the quicksilver liquid metal mercury, and this pungent, fuming red-brown substance. Based on the Greek word *bromos*, bromine's name means 'stench'. Corrosive and toxic, it is dubbed 'the Houdini element' since it is impossible to confine for any length of time. Electronegative like the rest of the halogen family, it eats plastic and rubber stoppers, and even attacks the normally incorruptible Teflon.

It is no surprise that something this reactive is not found in its elemental form on Earth. It was isolated by 24-year-old student Antoine-Jérôme Balard (1802–76) in 1826, from salty residues left after evaporating seawater (still the element's principal reservoir). German chemist Carl Jacob Löwig (1803–90) is also credited with discovering bromine independently. Until recently, it was ubiquitous in fungicides, pesticides, petrol additives and bleaching pool cleaners, but many bromine compounds are now regulated due to their damaging effect on the ozone layer.

C ⬤ ⬤ Br
Atomic radius: 94pm

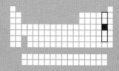

Group 17, Period 4
Halogen

Atomic number
35

Br

Bromine

79.90
Atomic weight

Melting point: -7.3°C (18.8°F)
Boiling point: 59°C (138°F)
Density: 3.122 g/cm³

Common isotopes: Br-79, Br-81
State (at STP): Gas
Colour: Red-brown

# Krypton

Krypton is 'the hidden one'. With no smell or colour, this noble gas was discovered in 1898 by William Ramsay and Morris Travers, along with neon and xenon, in air. It was revealed by its strong orange-red and green spectral lines. Once the major elements of the atmosphere – oxygen and nitrogen – were removed, fractional distillation of liquefied air (see page 106) separated the remaining minor components. Krypton exists in concentrations of about one part per million. However, since the gas is a natural product of uranium decay – and is heavy enough not to be lost to space – it is steadily accumulating.

Like the other noble gases, krypton emits light in an electrical discharge tube. Its brilliant white glow guides aircraft into land, even in thick fog. However, its expense prohibits wider use: at more than 8,000 times more scarce than argon in Earth's atmosphere, krypton costs about 100 times as much to extract. As Ramsay suspected, writing in 1902, krypton is not entirely inert and can be persuaded to take part in reactions.

C 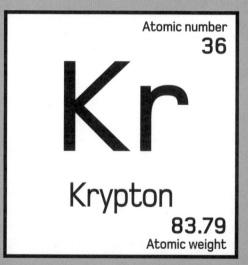 Kr
Atomic radius: 88pm

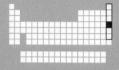

Atomic number
36

# Kr

## Krypton

83.79
Atomic weight

Melting point: −157.4°C (−251.2°F)
Boiling point: −153°C (−243°F)
Density: 0.003733 g/cm³

Common isotopes: Kr-84, Kr-86, Kr-82
State (at STP): Gas
Colour: Colourless

# Rubidium

An element often associated with colour, rubidium was the second of two elements discovered by Robert Bunsen and Gustav Kirchhoff in 1861, using their new technique of spectroscopy. They studied salts purified from the mineral lepidolite, burning them in a bunsen flame. The light produced was passed through a prism, splitting into its component colours and revealing rubidium's characteristic twin ruby spectral lines. The element itself, however, was not isolated until 1928.

Rubidium is a soft, silvery metal that is barely solid at room temperature. It is ferociously reactive (the alkali metals of group 1 get more volatile down the group) and combines powerfully and quickly with oxygen in the air. Its reaction with water is second only to that of caesium. Certain microwave frequencies make atoms of rubidium resonate with extreme regularity, allowing them to be used for timekeeping. Cheaper and more portable than caesium clocks, they are used in GPS satellites, and to control TV station and mobile phone frequencies.

C 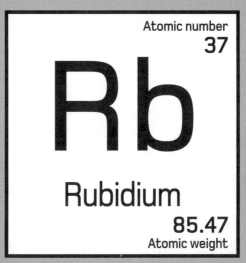 Rb

Atomic radius: 265pm

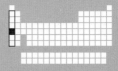

Group 1, Period 5
Alkali metal

Atomic number
37

# Rb

Rubidium

85.47
Atomic weight

Melting point: 39.3°C (102.8°F)
Boiling point: 688°C (1270°F)
Density: 1.532 g/cm³

Common isotopes: Rb-85, Rb-87
State (at STP): Solid
Colour: Silver

# Strontium

Calcium's closest chemical relative has a bad reputation, as no fewer than 26 of its 30 isotopes are radioactive. Because alkaline earth metal chemistry is markedly uniform, it's unsurprising that strontium incorporates easily into bones and teeth. Levels of radioactive strontium-90, produced by nuclear fission, increased dramatically in babies' milk teeth during the period of above-ground nuclear tests (1945 to 1963) – evidence that was instrumental in treaties banning such testing. However, fallout from the 1986 Chernobyl nuclear accident seeded vast areas of Europe with Sr-90.

Named after the Scottish town of Strontian where it was discovered in 1790, strontium is a soft, greyish, reactive metal. It is pyrophoric, meaning that it bursts into flame on contact with air. Elemental strontium does not have many commercial applications (although Sr-89 is a radioisotope used in some cancer therapies), but strontium carbonate gives a bright red colour to fireworks and distress flares.

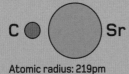

Atomic radius: 219pm

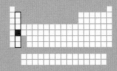

Group 2, Period 5
Alkaline earth metal

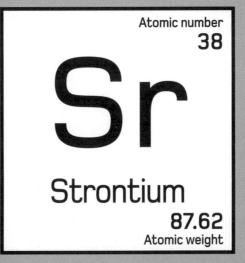

Atomic number
38

Sr

Strontium

87.62
Atomic weight

Melting point: 777°C (1431°F)
Boiling point: 1382°C (2520°F)
Density: 2.64 g/cm³

Common isotopes: Sr-88, Sr-86, Sr-87
State (at STP): Solid
Colour: Silver

# Yttrium

Although strictly a transition metal, yttrium's similarities in both chemistry and distribution to the lanthanoids mean it is also classed as a 'rare earth' element. Its 1794 discovery by Finnish chemist Johan Gadolin marked the beginning of an element bonanza: in 1843, Carl Mosander (1797–1858) found two more oxides – of terbium and erbium – lurking in its mineral ore and, like a matryoshka doll, these in turn contained nine more lanthanoids (see page 134). From the outset, this plain-looking metal's principal talent was evident: 'hiding' other lanthanoids. Yttrium only ever forms 3+ ions, meaning that yttrium oxide 'hosts' similarly sized rare earth metals in the same oxidation state.

Yttrium–aluminium garnet (YAG), a diamond-like artificial crystal, is doped with neodymium to make the most common solid-state laser, used in surgery, dentistry and for cutting metal. Another yttrium compound – yttrium barium copper oxide (YBCO) – was the first superconductor found to show zero electrical resistance at the relatively high temperature of -180.15°C (-292.27°F).

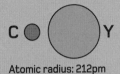

Atomic radius: 212pm

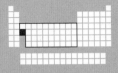

Group 3, Period 5
Transition metal

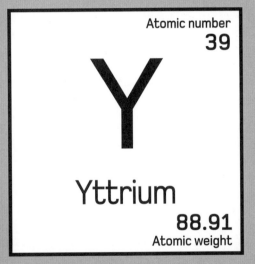

Atomic number

39

Y

Yttrium

88.91

Atomic weight

Melting point: 1526°C (2779°F)
Boiling point: 3336°C (6037°F)
Density: 4.469 g/cm³

Common isotopes: Y-89
State (at STP): Solid
Colour: Silver

# Zirconium

Chemically similar to titanium, zirconium is tough stuff. It is reasonably reactive and on exposure to air quickly forms a hard surface layer of zirconium dioxide that makes the metal almost impervious to chemical attack. It also withstands high temperatures and does not absorb neutrons, so it is used inside nuclear reactors and as cladding for nuclear fuels.

Zirconium dioxide, or zirconia, is a ceramic that is used to make chemically inert lab equipment, low-friction bearings and super-sharp knives. When stabilized by yttria (yttrium oxide), it forms cubic zirconia crystals. These synthetic crystals are almost as hard as diamond and just as good-looking. On Earth, zirconium is found principally as zircon, a silicate gemstone. Zircons are so hard they move unperturbed through the rock cycle, and can be used to date generations of continental rock long ago reworked by the Earth. Gritty zircon sand is so resistant to heat that it is used to line glass-making furnaces and to create giant ladles for scooping molten metal in industrial foundries.

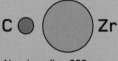

Atomic radius: 206pm

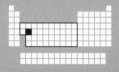

Group 4, Period 5
Transition metal

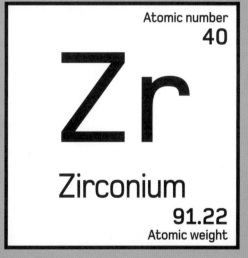

Atomic number
40

# Zr

Zirconium

91.22
Atomic weight

Melting point: 1855°C (3371°F)
Boiling point: 4409°C (7968°F)
Density: 6.506 g/cm³

Common isotopes: Zr-90, Zr-94, Zr-92
State (at STP): Solid
Colour: Silver

# Niobium

Element 41 has a tortured history. In 1801, Charles Hatchett (1765–1847) first identified the metal in its ore and named it columbium. Eight years later, however, William Hyde Wollaston (1766–1828) declared that columbium was in fact nothing but the already known tantalum. In 1846, German chemist Heinrich Rose (1795–1864) realized that tantalum ore contained two elements. Rose named the 'new' element niobium, after Niobe, the daughter of Tantalus in Greek mythology.

Niobe was a tragic figure who lost all her children as punishment for hubris, but the element hasn't been held back by such pessimistic associations. In fact, it has taken us to the Moon – surely the most exuberant expression of human optimism. This softish, grey ductile metal is one of five refractory metals (also molybdenum, tantalum, tungsten and rhenium) renowned for their heat resistance, and is used in nickel-based superalloys for jet engine parts and rocket nozzles. Niobium is also used in the superconducting coils of particle accelerators and MRI scanners.

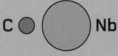

C Nb

Atomic radius: 198pm

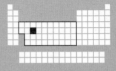

Group 5, Period 5
Transition metal

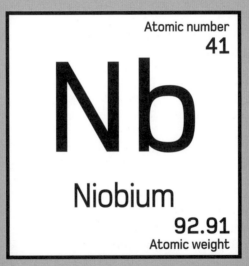

Atomic number
41

Nb

Niobium

92.91
Atomic weight

Melting point: 2477°C (4491°F)
Boiling point: 4744°C (8571°F)
Density: 8.57 g/cm³

Common isotopes: Nb-93
State (at STP): Solid
Colour: Grey

# Molybdenum

Molybdenum is a high performer. One of the five refractory metals, this silver-grey metal has an extreme resistance to heat that it shares with its near neighbours, niobium, tantalum, tungsten and rhenium. Adding molybdenum to steel has the same effect as adding tungsten, toughening up the metal and keeping it hard at high temperatures. Molybdenum steel is a high-speed steel (HSS) – hard-wearing and capable of drilling faster than standard high-carbon steel.

Despite its rarity, molybdenum is an important catalyst in both oil refining and biological systems. In bacterial enzymes, the element catalyzes the breaking of the nitrogen bond, enabling nitrogen gas in the atmosphere to be absorbed. Nitrogen-fixing plants, such as legumes, play host to these bacteria in their roots, providing a pathway for nitrogen to enter the biosphere. The metal is an essential trace element for nearly all living things – necessary to the function of dozens of enzymes. Even so, you won't need more than about 300 mg in a lifetime.

C ◯ ⬤ Mo

Atomic radius: 190pm

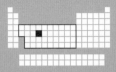

Group 6, Period 5
Transition metal

Atomic number
42

# Mo

Molybdenum

95.95
Atomic weight

Melting point: 2623°C (4753°F)
Boiling point: 4639°C (8382°F)
Density: 10.22 g/cm³

Common isotopes: Mo-98, Mo-96, Mo-95
State (at STP): Solid
Colour: Grey

# Technetium

With no stable isotopes, element 43 is the lightest radioactive element. Although it is produced naturally as a decay product of uranium, only about 0.2 nanograms are found in each kilogram (2.2 lb) of uranium ore. This explains the 68-year gap between Mendeleev's 1869 prediction of *eka-manganese* and its eventual isolation as technetium in 1937. Carlo Perrier and Emilio Segrè, working at the University of Palermo, in Sicily, identified the new element in molybdenum foil taken from inside the cyclotron particle accelerator at Berkeley, California. They arrived at the name from the Greek root *technetos*, meaning 'artificial'. The subsequent 1952 detection of technetium's absorption spectrum in red giant stars confirmed the theory of stellar nucleosynthesis (see page 90). Technetium-99, which decays by releasing a single photon of gamma radiation, is used for medical imaging and targeted treatment of cancers. It has a half-life of about six hours and does not interact with the body, since it does not occur in nature.

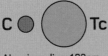

C ● ⬤ Tc

Atomic radius: 183pm

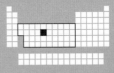

Group 7, Period 5
Transition metal

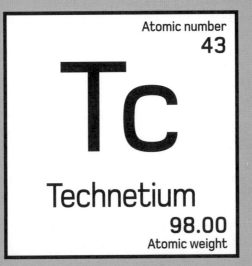

Atomic number
43

# Tc

Technetium

98.00
Atomic weight

Melting point: 2157°C (3915°F)
Boiling point: 4265°C (7709°F)
Density: 11.5 g/cm³

Common isotopes: None stable
State (at STP): Solid
Colour: Silver

# Ruthenium

The platinum group metals (PGMs) are a family of the most expensive elements on the periodic table. This group of six metals occurs as a block in the middle of the transition metals, and consists of ruthenium, rhodium, palladium, osmium, iridium and platinum. Their expense is due both to their rarity and to their usefulness as industrial catalysts. Ruthenium is a minor component of platinum ore – the PGMs tend to occur together in the same deposits, and most were isolated from their alloy, platina, at the turn of the 19th century. The discovery of ruthenium, however, lagged behind by some 50 years.

In its interactions with other elements, ruthenium plays a bit-part role. Effective in increasing the hardness of platinum and palladium alloys, it is used in wear-resistant electrical contacts. It is also a lesser ingredient in several high-performance superalloys. Indeed, ruthenium's most prominent use has been the 'RU' nib on the Parker 51 fountain pen: this 14-carat gold nib was tipped with 96.2 per cent ruthenium and 3.8 per cent iridium.

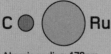 C ○ ● Ru

Atomic radius: 178pm

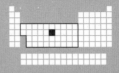

Group 8, Period 5
Transition metal

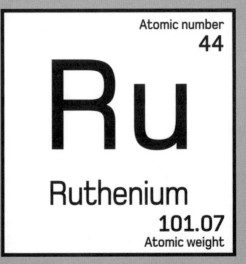

Atomic number
44

# Ru

Ruthenium

101.07
Atomic weight

Melting point: 2334°C (4233°F)
Boiling point: 4150°C (7502°F)
Density: 12.37 g/cm³

Common isotopes: Ru-102, Ru-104, Ru-101
State (at STP): Solid
Colour: Silver

# Rhodium

In 1979, the *Guinness Book of Records* presented Paul McCartney with a rhodium-plated record, recognizing him as the best-selling artist of all time. One of the rarest and most expensive metals on Earth, rhodium is even more valuable than platinum. Element 45 was the last PGM discovered by William Hyde Wollaston in 1804 – its name refers to the rose-coloured solutions he created with this noble metal.

Rhodium was first used as a corrosion-resistant plating, and is still used on some sterling silvers to increase tarnish resistance and shine. It is used to make thermocouple wires that measure temperatures up to 2,000°C (3,800°F) inside furnaces, and in artificial pacemakers, to make the exceedingly fine wires that pulse electrical signals direct into the cells of the heart. However, rhodium is mostly too rare and too expensive to be used on its own, and is often alloyed with platinum. Most of the small quantities of rhodium produced each year go into three-way catalytic converters that reduce harmful automobile emissions.

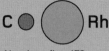

C ● ⬤ Rh

Atomic radius: 173pm

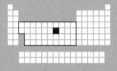

Group 9, Period 5
Transition metal

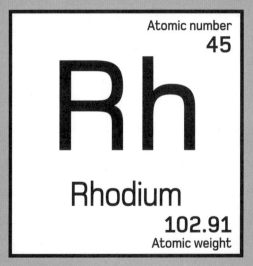

Atomic number
45

# Rh

Rhodium

102.91
Atomic weight

Melting point: 1964°C (3567°F)
Boiling point: 3695°C (6683°F)
Density: 12.41 g/cm³

Common isotopes: Rh-103
State (at STP): Solid
Colour: Silver

# Palladium

In 1802, William Hyde Wollaston's secret method for purifying platinum ore yielded a new element. Fearing that formal publication of his discovery might draw other chemists to the field and jeopardize his plans to monopolize the platinum market, he instead placed a newspaper ad and began to sell the metal through an agent as 'the new silver'. With this ploy, he hoped to establish priority and step in later to claim the glory. However, when others deemed the new metal a 'contemptible fraud,' Wollaston was forced to come clean, exposing his deceit and grievously damaging his reputation.

Wollaston named his silvery, hard and dense noble metal palladium, after the newly discovered asteroid Pallas. Extremely rare, it is among the most valuable of all metals, and also an important industrial catalyst. It is used in the 'cracking' reactions that convert crude oil compounds into useful petrochemicals, and in catalytic converters that reduce harmful automotive emissions. It is also used in electronic capacitors, dental fillings and jewellery.

C 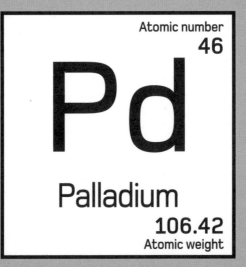 Pd

Atomic radius: 169pm

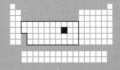

Group 10, Period 5
Transition metal

Atomic number
46

# Pd

Palladium

106.42
Atomic weight

Melting point: 1554.9°C (2830.8°F)
Boiling point: 2963°C (5365°F)
Density: 12.02 g/cm³

Common isotopes: Pd-106, Pd-108, Pd-105
State (at STP): Solid
Colour: Silver

# Silver

Lustrous silver is a precious metal, highly prized since antiquity. It is found in its unalloyed native state, and in several sulfide minerals and lead ores. More reactive than gold, silver tarnishes slowly in air and pure nuggets are rare. It is used for jewellery and ornaments, and alongside copper and gold is a coinage metal (see page 132), minted as currency by many cultures. Silver from the colonial New World bankrolled the Spanish Empire: the Spanish silver dollar was accepted currency across the 'civilized' world, though today it is probably best remembered as the 'pieces of eight' so coveted by pirates.

Despite these ancient associations, silver has many modern uses. As the best practical conductor of electricity and heat, it is used in high-end audio equipment (though its tendency to corrode means that gold is often used as a substitute). It also makes super-reflective mirrors used in solar refectors and telescope mirrors. What's more, microbes cannot grow on silver, so it is used for sterile dressings, medical equipment and to purify water.

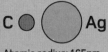

C  Ag

Atomic radius: 165pm

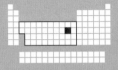

Group 11, Period 5
Transition metal

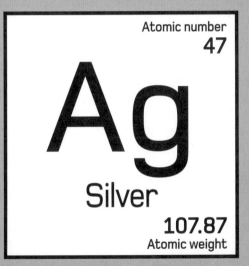

Atomic number
47

Ag

Silver

107.87
Atomic weight

Melting point: 961.8°C (1763.2°F)
Boiling point: 2162°C (3924°F)
Density: 10.501 g/cm³

Common isotopes: Ag-107, Ag-109
State (at STP): Solid
Colour: Silver

# Cadmium

Cadmium is the artist's favourite element. This silver-blue metal is unremarkable until it partners up with other elements. Most transition metals create colourful compounds, but those formed with cadmium have vibrancy and punch unlike any other. From the late 1830s, a new range of light-resistant cadmium-based yellow, orange and red pigments burst onto the market and, despite their cost were rapidly adopted. Claude Monet's *Autumn at Argenteuil* and *Water-Lilies* are particularly associated with bright cadmium pigments.

Cadmium compounds are also used to colour plastics (the orange colour of gas-main piping is cadmium sulfoselenite), but use in children's products is prohibited: cadmium is a heavy metal that is toxic even in small quantities. Chemically similar to zinc, element 48 is ubiquitous in zinc deposits, and present in trace amounts in every piece of galvanized steel. The heavy metal's chief use, however, is in making nickel–cadmium rechargeable batteries, building a legacy of toxic waste for the future.

C 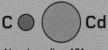 Cd

Atomic radius: 161pm

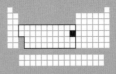

Group 12, Period 5
Transition metal

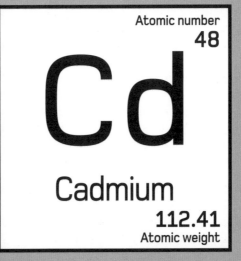

Atomic number
48

# Cd

Cadmium

112.41
Atomic weight

Melting point: 321.1°C (609.9°F)
Boiling point: 767°C (1413°F)
Density: 8.69 g/cm³

Common isotopes: Cd-114, Cd-112, Cd-111
State (at STP): Solid
Colour: Silver-blue

# Indium

Like cadmium, indium is produced mainly as a by-product of zinc ore processing. Slightly more abundant than silver or mercury, it is nevertheless a relatively rare metal. Soft and malleable with a low melting point, shiny indium ignites easily, giving off a violet flame that shows the distinctive indigo-blue line in its spectrum that gave the metal its name.

Indium is alloyed to other metals to lower their melting points. Its stickiness, and ability to bind to other metals and itself, sees it used for solders and sealing rings. Unlike many metals, indium remains workable at low temperatures, and is therefore useful in cryogenic pumps and high-vacuum environments. Indium tin oxide (ITO) is a tough, transparent, conductive oxide, used on LCD screens and solar cells. A liquid crystal display needs an electric potential across each cell, but also needs to be see-through: when you watch a flatscreen TV, you are watching through a thin film of ITO. Coating windows with ITO, meanwhile, reflects infrared light (heat) and allows them to be de-iced by an electric current.

C ⬤ ◯ In

Atomic radius: 156pm

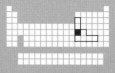

Group 13, Period 5
Post-transition metal

Atomic number
49

In

Indium

114.82
Atomic weight

Melting point: 156.6°C (313.9°F)
Boiling point: 2072°C (3762°F)
Density: 7.31 g/cm³

Common isotopes: In-115, In-113
State (at STP): Solid
Colour: Silver

# Tin

Tin is a somewhat mystical metal, with associations spinning back to our prehistoric past: tin was the element that gave rise to the Bronze Age. In fact, it was so crucial to the Roman Empire that the distant Cornish tin mines of Britain became a strategically important resource. Its chemical symbol, Sn, comes from its Latin name, *stannum*.

Tin is soft, easily melted and cast. It forms a shiny silvery-white, crystalline surface. When bent, the twinned crystals shear with a high-pitched creak, known as the 'tin cry'. Tin gives an attractive corrosion-resistant plating for steel, which is also non-toxic and used to coat food containers. An alloy metal with few rivals, it is also used in solder, although 'tin pest' can be a problem: in cold temperatures, metallic white tin (aka 'beta tin') changes to a flaky, non-metallic allotrope called 'alpha tin', which causes deterioration and ruins soldered electrical contacts. Niobium—tin alloy, meanwhile, is a high-temperature superconductor, used in the coils of superconducting magnets.

C  Sn

Atomic radius: 145pm

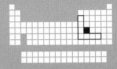

Group 14, Period 5
Post-transition metal

Atomic number
50

# Sn

Tin

118.71
Atomic weight

Melting point: 231.9°C (449.5°F)
Boiling point: 2602°C (4716°F)
Density: 7.287 g/cm³

Common isotopes: Sn-120, Sn-118, Sn-116
State (at STP): Solid
Colour: Silver-white

# Antimony

Antimony is another ancient, alchemical element. This metalloid was known to the Ancient Egyptians, who crushed the mineral stibnite into a black powder, to use for their heavy eye make-up. Stibnite – antimony's principal ore – is also the origin of its chemical symbol. Roman texts describing 'feminine antimony' are interpreted as referring to the metallic form, showing they had a method for isolating the element. Certainly, by the 16th century, recipes for producing metallic antimony were widely available. Followers of the physician-philosopher Paracelsus advocated the health benefits of ingesting antimony. The fact that it caused vomiting due to its toxicity was the whole point, since it was supposed to purge the body.

Sometimes called a 'poor metal', antimony is not a good conductor of electricity and heat. Even so, this soft, brilliantly shiny material is mainly used in tin and lead alloys, such as pewter. Most antimony is used to harden lead in the alloy plates used in lead-acid batteries.

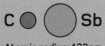
C Sb

Atomic radius: 133pm

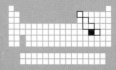

Group 15, Period 5
Metalloid

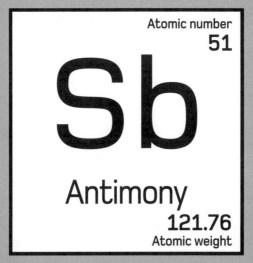

Atomic number
51

# Sb

Antimony

121.76
Atomic weight

Melting point: 630.6°C (1167.1°F)
Boiling point: 1587°C (2889°F)
Density: 6.685 g/cm³

Common isotopes: Sb-121, Sb-123
State (at STP): Solid
Colour: Silver

# Tellurium

The rare element tellurium is notable as one of the only elements that forms compounds with gold. Gold is generally far too noble to associate with any of its neighbours on the periodic table, but this metalloid was first identified in 1782 in gold ore from Transylvania, and was initally dubbed *aurum paradoxium*. Rather neatly – but completely coincidentally – tellurium's abundance is also comparable to that of gold. Martin Heinrich Klaproth named it from a Latin word for the ground in 1798. Tellurium is a brittle, silver-white metalloid that forms large crystals when allowed to cool from a molten liquid. Chemists who work with it are best avoided: contact with even small quantities gives the handler a garlicky body odour that can last for weeks. Every dog has its day, however, and cadmium telluride solar cells are currently the best available. CDs, DVDs and Blu-ray discs all encode data on a thin layer of a 'phase-changing' silver, indium, tin and tellurium compound. Laser heating turns small blobs of this crystalline material into glass, altering its optical properties and allowing information to be stored.

C ◯ Te
Atomic radius: 123pm

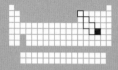

Group 16, Period 5
Metalloid

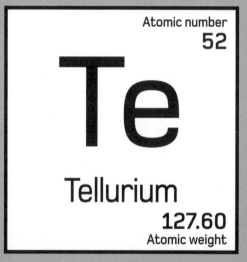

Atomic number
52

# Te

Tellurium

127.60
Atomic weight

Melting point: 449.5°C (841.1°F)
Boiling point: 988°C (1810°F)
Density: 6.232 g/cm³

Common isotopes: Te-130, Te-128, Te-126
State (at STP): Solid
Colour: Silver

# Iodine

Iodine is the first of the solid halogen elements, although it acts as if not entirely convinced of this fact. When heated, it 'sublimates' – turns from the solid phase directly into the gas phase, without pausing to become a liquid. While most elemental solids are shiny silver metals or dull grey-black non-metals (with the notable exceptions of copper, gold and sulfur), iodine is blue-black. As a gas it is a surprisingly vivid violet.

The reactivity of the halogens decreases down the group (see page 148). Accordingly, iodine is less reactive than either chlorine or bromine, but more reactive than astatine – in practice, however, astatine is so rare that iodine may as well be the least reactive halogen. As a powerful antimicrobial, it is used to purify water and as an antiseptic: iodine is the active ingredient in the yellow-brown paste painted onto patients before surgery. It is also the heaviest essential element widely used in biology. A deficiency in the diet causes hypothyroidism – the inability to properly regulate the metabolism.

C 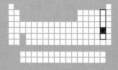 I

Atomic radius: 115pm

Group 17, Period 5
Halogen

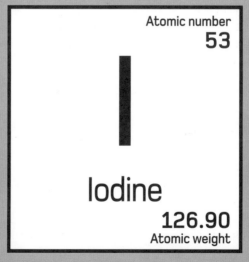

Atomic number
53

I

Iodine

126.90
Atomic weight

Melting point: 113.7°C (236.7°F)
Boiling point: 184.3°C (363.7°F)
Density: 4.93 g/cm³

Common isotopes: I-127
State (at STP): Solid
Colour: Blue-black

# Xenon

With a name meaning 'the strange one', xenon was discovered in 1898 at the same time as neon and krypton by William Ramsay (see page 240). Even scarcer than krypton, this colourless, odourless noble gas is the smallest fraction of air, found as a monatomic gas in vanishingly small proportions of just 90 parts per billion.

For such a rare and expensive gas, however, xenon is widely used. In a discharge tube, or under a high-pressure electrical arc, the gas flashes a bright, white light – a property used in electronic camera tube flashes, car headlamps and cinema projector lamps. Heavy, but chemically inert, xenon is also used for futuristic ion thrusters for spacecraft. It was also the first of the noble gases to break ranks and form a compound: Neil Bartlett's 1962 synthesis of xenon hexafluoroplatinate proved that the group 18 elements weren't perhaps so 'noble' after all. Xenon was also at the forefront of atomic force microscopy in 1981, when Don Eigler used a probe tip to spell out the letters 'IBM' in 35 xenon atoms.

C 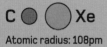 Xe

Atomic radius: 108pm

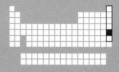

Group 18, Period 5
Noble gas

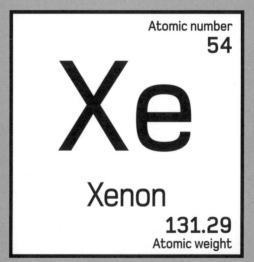

Atomic number
54

# Xe

Xenon

131.29
Atomic weight

Melting point: −111.8°C (−169.2°F)
Boiling point: −108.1°C (−163°F)
Density: 0.005887 g/cm³

Common isotopes: Xe−132, Xe−129, Xe−131
State (at STP): Gas
Colour: Colourless

# Caesium

Only ever safely handled under oil, or in an inert argon or nitrogen atmosphere, caesium is a pale gold-coloured metal with a fiery temper. It has the largest atoms of any element, so removing its lone outer electron is simplicity itself. This alkali metal is soft and would melt in the hand if it were safe to attempt such a test. But caesium, like a loaded gun with a hair trigger, is primed to go off at the slightest touch of air or water. In fact, most chemists consider caesium to be the most reactive element of all.

Within the Earth's crust, caesium is an 'incompatible element': its large size prevents its atoms from incorporating into most solid crystal lattices. Thus, its only source is rare minerals in pegmatite rocks that crystallize last from magmas. Caesium also underpins global standards for time: the world's most accurate atomic clocks – accurate to 1 second in over a million years – measure the finely tuned frequency of microwave light absorbed or emitted by caesium-133 atoms. One second is defined precisely as the duration of 9,192,631,770 such cycles.

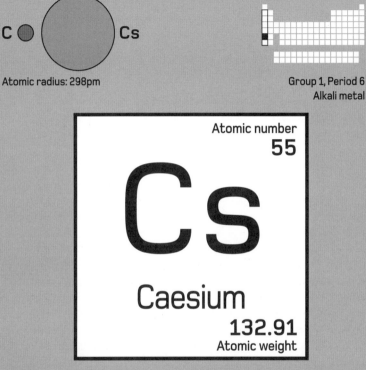

C ● Cs

Atomic radius: 298pm

Group 1, Period 6
Alkali metal

Atomic number
55

Cs

Caesium

132.91
Atomic weight

Melting point: 28.4°C (83.2°F)
Boiling point: 671°C (1240°F)
Density: 1.873 g/cm³

Common isotopes: Cs-133
State (at STP): Solid
Colour: Pale gold

# Barium

Heaviness is the quality that most defines barium. In the lower reaches of the periodic table, high density and atomic weight become normal. Barium was first isolated by Humphry Davy in 1808, by electrolysis of molten barium salts, but Carl Wilhelm Scheele had identified a new element in the mineral barite as early as 1774. More reactive than the lighter alkaline earth metals, soft silvery barium is not found uncombined in nature: it is reactive enough to be used as a 'getter' in vacuum tubes, removing even the last remaining traces of gases.

Crushed barite (barium sulfate), also known as heavy spar, is used as a weighting agent in drilling mud, providing pressure when drilling oil and gas wells. Barium sulfate is also the sickly white fluid used in 'barium meals': when ingested, the heavy compound blocks X-rays and allows the large intestine or oesophagus to be imaged. Barium sulfate is so insoluble that it is safe to ingest, but barium carbonate, a rat poison, dissolves in stomach acid and is fatal in doses of less than 1 gram.

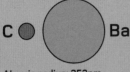

C ● ● Ba

Atomic radius: 253pm

Group 2, Period 6
Alkaline earth metal

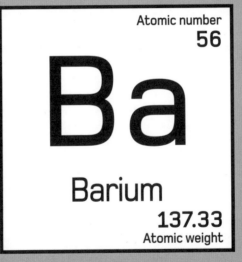

Atomic number
56

# Ba

Barium

137.33
Atomic weight

Melting point: 727°C (1341°F)
Boiling point: 1897°C (3447°F)
Density: 3.594 g/cm³

Common isotopes: Ba-138, Ba-137, Ba-136
State (at STP): Solid
Colour: Silver

# Lanthanum

Lanthanum is the first of the eponymous lanthanoid series that usually hang suspended beneath the bulk of the periodic table: a full-length version of the table would show where these 'f-block' elements really fit in (see page 134). Discovered in 1839 in the mineral cerite, by Swedish chemist Carl Gustaf Mosander, it was named after the Greek *lanthano*, meaning 'to lie hidden'. In 1842, Mosander was able to refine lanthanum further and found another element lurking inside, which he named didymium (see page 286).

Like most lanthanoids, lanthanum is a soft, ductile, shiny metal. A typical hybrid car contains 10–15 kilograms (22–33 lb) of the metal in the electrodes of its batteries. Although not particularly scarce, lanthanum is expensive to produce, so the anodes are not pure and instead contain about 50 per cent of the rare earth metal. Electron microscope probes are made with tips of lanthanum hexaboride, while lanthanum oxide added to high-quality camera lenses increases hardness and clarity.

C 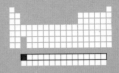 La

Atomic radius: Unmeasured

Period 6
Lanthanoid

Atomic number
57

La

Lanthanum
138.91
Atomic weight

Melting point: 920°C (1688°F)
Boiling point: 3464°C (6267°F)
Density: 6.145 g/cm³

Common isotopes: La-139, La-138
State (at STP): Solid
Colour: Silver

# Cerium

The story of the lanthanoids really begins with cerium. While Johan Gadolin had identified a new element in the mineral yttria from Ytterby (see page 246), in 1794, another heavy mineral – ceria – was attracting the interest of Wilhelm Hisinger (1766–1852), Jöns Jakob Berzelius and Martin Klaproth. In 1839, Carl Gustaf Mosander separated ceria into pure cerium and another 'earth', that he called lanthana. This in turn yielded an entire 'nested series' of new rare earth elements.

Named after the large asteroid Ceres, cerium is a pyrophoric ('fire-bearing') element: tiny flakes ignite spontaneously on contact with air. Striking a 'mischmetal' firesteel against ridged steel chips off small particles that can light campfires and cigarette lighters. This metal mix, usually made of cerium, lanthanum and smaller amounts of neodymium and praseodymium, also creates spectacular showers of sparks for movie special effects. Cerium oxide, meanwhile, is added to glass, giving it a golden-yellow colour and serving to block UV rays.

C ◯ Ce

Atomic radius: Unmeasured

Period 6
Lanthanoid

Atomic number
58

Ce

Cerium

140.12
Atomic weight

Melting point: 795°C (1463°F)
Boiling point: 3443°C (6229°F)
Density: 6.77 g/cm³

Common isotopes: Ce-140, Ce-142, Ce-138
State (at STP): Solid
Colour: Silver

# Praseodymium

**P**raseodymium is one of a pair of rare earth 'twins' discovered at the same time. In 1841, Swedish chemist Carl Mosander separated his 'lanthana' mineral into lanthanum and another element he named didymium. However, like a Russian doll, didymium was split and split again to reveal no less then five new elements inside. In 1885, Austrian chemist Carl Auer von Welsbach (1858–1929) found 'praseodidymium' ('green twin') and 'neodidymium' ('new twin') in didymium.

The name didymium is still is use for mixtures of praseodymium and neodymium. 'Didymium glass' is used in glassblowers' and blacksmiths' goggles. These safety glasses effectively filter out the yellow glare emitted by hot sodium, allowing much better vision of the work in progress, while protecting eyes from damage. Soft, silvery and easily workable, praseodymium is fairly reactive, quickly forming a greenish oxide tarnish on its surface. It is a minor component of mischmetal 'flints' (see page 284) and tints cubic zirconia green to mimic precious peridot.

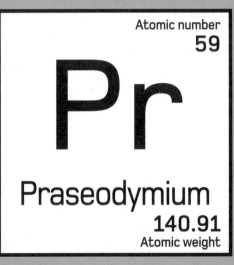

C ◉ Pr

Atomic radius: 247pm

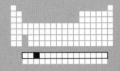

Period 6
Lanthanoid

Atomic number
59

Pr

Praseodymium
140.91
Atomic weight

Melting point: 935°C (1715°F)
Boiling point: 3520°C (6368°F)
Density: 6.773 g/cm³

Common isotopes: Pr-141
State (at STP): Solid
Colour: Silver

# Neodymium

Neodymium is one of the elements that has enabled the revolutionary miniaturization of technology. Element 60 (the 'new twin', see page 286) is used to make widely available 'neodymium magnets' (in fact, an iron and boron compound $Nd_2Fe_{14}B$). These small permanent magnets are incredibly powerful: a neodymium magnet weighing just a few grams can lift a thousand times its own weight. They are ubiquitous wherever strong, but light magnets are required – in laptop computers, hard disk drives, mobile phone vibrators and speakers, headphones, microphones, drone 'multicopters', DIY drills and impact drivers and even some wind turbines.

Yttrium–aluminium–garnet crystals doped with neodymium make the most common solid-state 'Nd:YAG' lasers – used to cut and weld body tissues, teeth and sheet steel. Like many rare earth elements, neodymium is also used in glass staining, giving a range of soft pink to lilac hues. As the second-most abundant lanthanoid element after cerium, it is not particularly rare.

C 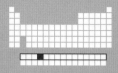 Nd

Atomic radius: 243pm

Period 6
Lanthanoid

Atomic number
60

# Nd

Neodymium

144.24
Atomic weight

Melting point: 1024°C (1875°F)
Boiling point: 3074°C (5565°F)
Density: 7.007 g/cm³

Common isotopes: Nd-142, Nd-144, Nd-146
State (at STP): Solid
Colour: Silver

# Promethium

In Greek mythology, Prometheus was the Titan who stole fire from Mount Olympus and brought it to humanity. It earned him the singular and extraordinarily awful punishment of having his liver pecked out daily by an eagle, only to regrow overnight. No such retribution was inflicted on the discoverers of element 61, promethium – Chien Shiung Wu, Emilio Segrè and Hans Bethe.

As one of only two elements below atomic number 83 with no stable isotopes, promethium was the last lanthanoid to be discovered. It was synthesized at Oak Ridge National Laboratory, USA, in 1942, filling the last remaining gap in the periodic table. Promethium does not occur naturally on Earth, but 370 light-years away in space a curious star called Przybylski's Star pumps out promethium, technetium and many other heavy elements. Pm-147 is not the most stable isotope of promethium, but it has the most applications. This beta-emitter is used as fuel in radioisotope thermal generators (RTGs) that power spacecraft, and also for regulating the thickness of industrial sheet materials.

C  Pm

Atomic radius: 205pm

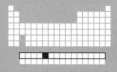

Period 6
Lanthanoid

Atomic number
61

# Pm

Promethium

145.00
Atomic weight

Melting point: 1042°C (1908°F)
Boiling point: 3000°C (5432°F)
Density: 7.26 g/cm³

Common isotopes: None stable
State (at STP): Solid
Colour: Silver

# Samarium

**N**ot many chemical elements are named after people, and those that do bear mostly the names of celebrated scientists. Samarium, however, holds the distinction of carrying the name of a mid-level military administrator. The element was discovered in 1879 by Paul-Émile Lecoq de Boisbaudran in the mineral samarskite, already named after Russian mine official Colonel Vasili Samarsky-Bykhovets. Hence the colonel became the very first person to have an element named after him.

After europium and ytterbium, samarium is the most reactive lanthanoid. Despite its classification as a rare earth metal, it is not in short supply – samarium is more common in the Earth's crust than tin. Samarium–cobalt magnets were the first rare earth magnets: now largely superseded by the lighter and stronger neodymium magnets (see page 288), they nevertheless still find plenty of use as pole pieces in electric guitar and bass pick-ups. Many samarium compounds are important catalysts for reactions in the plastics and petrochemical industries.

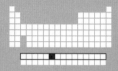

C  Sm

Atomic radius: 238pm

Period 6
Lanthanoid

Atomic number
62

# Sm

Samarium

150.36
Atomic weight

Melting point: 1072°C (1962°F)
Boiling point: 1794°C (3261°F)
Density: 7.52 g/cm³

Common isotopes: Sm–152, Sm–154, Sm–147
State (at STP): Solid
Colour: Silver

# Europium

On 1 January 2002, a new currency began to circulate in the European Union. Euro notes had a raft of high-tech anti-counterfeiting measures, designed to foil forgers. Key among these were fluorescent compounds picking out the stars and map of Europe, as well as decorative features, such as the Pont du Gard on the 5 Euro bill. These fluoresced brightly in red, green and blue under specific frequencies of light, and when analyzed by researchers at the University of Twente, Holland, were found to be metal complexes of element 63. With the name europium, and symbol Eu, this was an elegant elemental pun.

Europium oxide is used in 'phosphors' – powdered coatings on the inside of fluorescent lamps. These give compact fluorescent bulbs a more daylight quality. Unusually among the so-called 'rare earths', europium *is* actually fairly rare. The metal is found in many rare earth minerals, but only in proportions of around 0.1 per cent. According to some researchers, it could also have a shining future in memory chips for next-generation quantum computers.

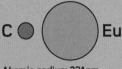

C ◯ Eu

Atomic radius: 231pm

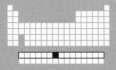

Period 6
Lanthanoid

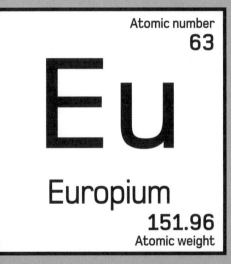

Atomic number
63

# Eu

Europium

151.96
Atomic weight

Melting point: 826°C (1519°F)
Boiling point: 1529°C (2784°F)
Density: 5.243 g/cm³

Common isotopes: Eu–153, Eu–151
State (at STP): Solid
Colour: Silver

# Gadolinium

Gadolinium is another lanthanoid that belongs to Paul-Émile Lecoq de Boisbaudran. Although identified in 1880 by the Swiss chemist Jean Charles Galissard de Marignac (1817–94) from its spectra, element 64 was isolated by the Frenchman from its mineral ore, gadolinite, in 1886.

The lanthanoids are all similar in size, and also all preferentially carry a 3+ charge: this means they have very similar chemistry and explains why they were so hard to separate. Lanthanoid compounds are toxic; however, chelation – surrounding a metal ion with organic molecules – makes them more stable and decreases their toxicity. Chelated gadolinium is used as a contrast agent for magnetic resonance imaging (MRI): injected into the bloodstream, its paramagnetism (the ability to 'boost' an applied external magnetic field) clearly marks its passage through soft tissue. Gadolinium is also effective at mopping up neutrons – it is used in nuclear reactor shielding, control rods to regulate fission reactions, and even to execute emergency reactor shutdowns.

C ◯ Gd
Atomic radius: 233pm

Period 6
Lanthanoid

Atomic number
64

Gd

Gadolinium
157.25
Atomic weight

Melting point: 1312°C (2394°F)
Boiling point: 3273°C (5923°F)
Density: 7.895 g/cm³

Common isotopes: Gd-158, Gd-160, Gd-156
State (at STP): Solid
Colour: Silver

# Terbium

A year after separating the mineral ceria into cerium, lanthanum and the 'earth' didymia in 1842, Carl Mosander set about another rare earth mineral, yttria. He split this into three fractions: the yellow-coloured earth he named erbia, the rose-coloured one terbia, and he kept the name yttria for the colourless one. However, the invention of the spectroscope in 1860 (see page 104) showed that there were more elements hidden inside erbia and terbia. Somehow, Mosander's names were swapped, and so in 1880 when Jean Charles Galissard de Marignac split Mosander's erbia, it revealed a new element called terbium.

Like europium, terbium's main use is in 'phosphors': terbium oxides fluoresce in yellow-green. Combined with blue europium (2+) and red europium (3+) phosphors, they create the daylight-mimicking trichromatic light, used in most ordinary fluorescent bulbs. What's more, some terbium compounds are 'triboluminescent', fluorescing when struck or put under stress: they may be able to reveal stresses building up in aircraft wings, buildings and structures.

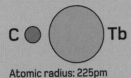

C ⬤ Tb

Atomic radius: 225pm

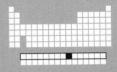

Period 6
Lanthanoid

Atomic number
65

# Tb

Terbium

158.93
Atomic weight

Melting point: 1356°C (2473°F)
Boiling point: 3230°C (5846°F)
Density: 8.229 g/cm³

Common isotopes: Tb-159
State (at STP): Solid
Colour: Silver

# Dysprosium

**D**ysprosium is another lanthanoid that came from the mineral ore erbia (see page 298). Despite its mythological-sounding name, the element's etymology is rather more pedestrian: over the year of 1886, Paul-Émile Lecoq de Boisbaudran carried out more than 30 attempts to isolate dysprosium from its oxide and so, on succeeding, he named it after the Greek word *dysprositos*, meaning 'hard to obtain'.

Modern extraction methods are fortunately slicker, and today dysprosium is something of a utility element. What it lacks in individuality, element 66 makes up for in team spirit: dysprosium shares the common chemical or physical properties of the lanthanoids and, as such, plays an accessory role in many rare earth compounds. Its high magnetic strength makes it perfect for data storage in computer hard disks, and it also makes up 6 per cent of neodymium magnets. Dysprosium, along with terbium and iron, also makes the 'magnetostrictive' alloy Terfenol-D – a material that changes shape in response to magnetic fields.

C 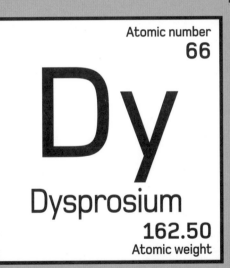 Dy

Atomic radius: 228pm

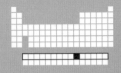

Period 6
Lanthanoid

Atomic number
66

# Dy

Dysprosium

162.50
Atomic weight

Melting point: 1407°C (2565°F)
Boiling point: 2567°C (4653°F)
Density: 8.55 g/cm³

Common isotopes: Dy-164, Dy-162, Dy-163
State (at STP): Solid
Colour: Silver

# Holmium

In 1878, Swiss chemists Marc Delafontaine (1837–1911) and Jacques-Louis Soret (1827–90) discovered some previously unrecorded spectral lines. With dramatic flair, they called their new discovery 'Element X'; a year later, Swedish chemist Per Teodor Cleve (1840–1905) separated it as holmium, naming it for the Latin name of his native city Stockholm.

Often overlooked, holmium has a claim to be 'the most underused element' in proportion to its abundance: while rare for a lanthanoid, it is still about 20 times more common than silver in the Earth's crust. Its chief uses are in scientific applications: holmium has the strongest magnetic moment (force) of all elements, and is used for field-boosting pole pieces in powerful low-temperature magnets. Added to glass, it makes filters for calibrating spectrometers. Holmium-doped YAG crystals, meanwhile, generate microwave-frequency laser light, used in some types of surgery because it is strongly absorbed by water and therefore does not penetrate far into tissue.

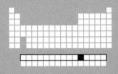

C ⬤ Ho

Atomic radius: Unmeasured

Period 6
Lanthanoid

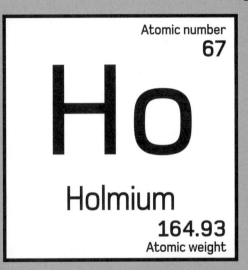

Atomic number
67

Ho

Holmium

164.93
Atomic weight

Melting point: 1461°C (2662°F)
Boiling point: 2720°C (4928°F)
Density: 8.795 g/cm³

Common isotopes: Ho-165
State (at STP): Solid
Colour: Silver

# Erbium

Soft, silvery and reactive with oxygen and water, erbium is just like any other lanthanoid metal. It was first identified in the earth terbia by Carl Mosander in 1843 (see page 298), but it was only in 1879 that Swedish chemist Per Teodor Cleve finally isolated the element.

Erbium helps connect the world: fibre-optic cables carry digital data flashing round the internet, but the beams of light reflecting back and forth within them lose energy, so short sections of erbium-doped glass fibre every 50 kilometres (30 miles) or so act as amplifiers, doubling the intensity of the light signal. When erbium atoms in the glass are excited, they emit light at the specific frequency that is least attenuated inside a fibre optic. Erbium compounds are often a delicate light pink – erbium trichloride is added to glass to colour jewellery and sunglasses. Erbium-doped Er:YAG crystals, meanwhile, produce near-infrared laser light, which is used for laser depilation, dentistry and dermatological surgery.

C 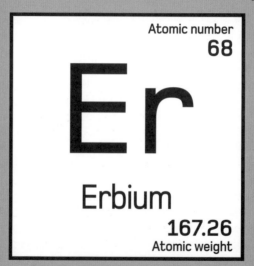 Er

Atomic radius: 226pm

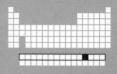

Period 6
Lanthanoid

Atomic number
68

# Er

Erbium

167.26
Atomic weight

Melting point: 1529°C (2784°F)
Boiling point: 2868°C (5194°F)
Density: 9.066 g/cm³

Common isotopes: Er-166, Er-168, Er-167
State (at STP): Solid
Colour: Silver

# Thulium

Thulium was one of seven lanthanoids to emerge from the rare earth mineral, erbia. Sweden's Per Teodor Cleve isolated its earth in 1879, naming it thulia, after *Thule*, the Greek name (or so he thought) for Scandinavia. In fact, Thule refers to 'the uttermost north' – the edges of the known world – but the name stuck and it was too late to change.

The difficulty of separating it, coupled with its scarcity (the rarest lanthanoid with the exception of promethium), makes thulium expensive. In 1911, English chemist Charles James (1880–1928) famously performed 15,000 recrystallizations to isolate a pure thulium chloride. Lanthanoids are difficult to separate because their comparable sizes and equal 3+ charge means they mix happily with one another. Elements mostly increase in size with atomic number thanks to electron shielding (see page 156), but the atomic radii of the lanthanoids actually decrease along the series. This 'lanthanoid contraction' affects f-block elements and is due to the poor shielding of electrons in the 4f subshell.

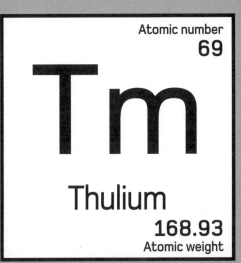

C ◯ Tm

Atomic radius: 222pm

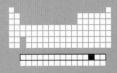

Period 6
Lanthanoid

Atomic number
69

Tm

Thulium

168.93
Atomic weight

Melting point: 1545°C (2813°F)
Boiling point: 1950°C (3542°F)
Density: 9.321 g/cm³

Common isotopes: Tm-169
State (at STP): Solid
Colour: Silver

# Ytterbium

In 1878, Jean Charles Galissard de Marignac discovered what would prove to be the last earth compound 'hidden' inside erbia. He called it ytterbia – another variation on the theme of Ytterby, Sweden, where so many of these rare earths originated. Ytterbia was split once again into two fractions in 1907 to reveal lutetium (see page 310), but isolation of pure ytterbium metal would have to wait until 1953.

Ytterbium makes the world's most accurate atomic clocks: some 100 times more accurate than the caesium clock, they can be used to measure the subtle effects of general relativity, such as the slowing of time due to gravity. 'Ticking' at around 518 *trillion* times per second, the hyperfine transitions within ytterbium atoms happen in optical wavelengths and are stimulated by lasers, rather than microwave radiation. Ytterbium radioisotopes can be used as a radiation source for mobile X-ray machines, while doping stainless steel with ytterbium has the unexpected effect of making the steel harder by reducing the size of metal grains.

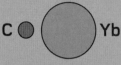

C ● ○ Yb

Atomic radius: 222pm

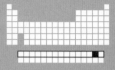

Period 6
Lanthanoid

Atomic number
70

# Yb

Ytterbium

173.05
Atomic weight

Melting point: 824°C (1515°F)
Boiling point: 1196°C (2185°F)
Density: 6.965 g/cm³

Common isotopes: Yb-174, Yb-172, Yb-173
State (at STP): Solid
Colour: Silver

# Lutetium

In 1907, three researchers independently discovered another hidden rare earth element inside the earth ytterbia. The French chemist Georges Urbain (1872–1938) named it 'lutecium', after *Lutetia*, the Latin name for his home city of Paris; Austrian scientist Carl Auer von Welsbach called it cassiopeium, after the constellation. Despite the dispute being officially settled as early as 1909, many German scientists patriotically continued to use von Welsbach's name until the 1950s. (The third man – Charles James, an Englishman working at the University of New Hampshire – kept out of the argument).

Lutetium is the hardest, densest rare earth metal, with the highest melting point. It is found principally in deposits of the phosphate mineral monazite, but the difficulty of extraction makes it prohibitively expensive and limits its use. Despite this, it is sometimes used as a catalyst in petroleum cracking. Although rare earth minerals including Lutetium are found around the globe, China dominates current production.

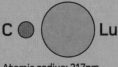

C ◯ Lu

Atomic radius: 217pm

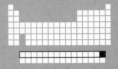

Period 6
Lanthanoid

Atomic number
71

# Lu

Lutetium

174.97
Atomic weight

Melting point: 1652°C (3006°F)
Boiling point: 3402°C (6156°F)
Density: 9.84 g/cm³

Common isotopes: Lu–175, Lu–176
State (at STP): Solid
Colour: Silver

# Hafnium

Dmitri Mendeleev's predictions for heavier elements did not pan out nearly as well as his predictions for lighter ones, but his 1869 prophecy of a heavier analogue for titanium and zirconium was one that came good. Lying on the boundary between the transitions and lanthanoids, researchers were initially unsure of where to hunt for it. In 1921, two young Danish scientists acting on a tip-off from Niels Bohr went looking in zirconium ores. They found it in short order.

Like zirconium, hafnium is exceptionally heat and corrosion resistant. But unlike its family member, hafnium strongly absorbs neutrons. For this reason it is used in control rods to regulate the fission inside a nuclear reactor (and it must be eliminated from zirconium before *that* metal can be predictably used in reactors). Currently, hafnium's rarity limits its use, but it may be one to watch for the future as it plays a key role in some superalloys: hafnium-tungsten carbide has the highest melting point of any known compound: 4,125°C (7,457°F).

C 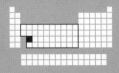 Hf

Atomic radius: 208pm

Group 4, Period 6
Transition metal

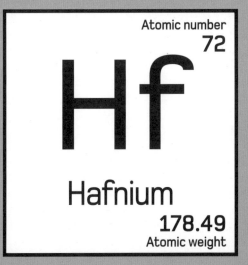

Atomic number
72

Hf

Hafnium

178.49
Atomic weight

Melting point: 2233°C (4051°F)
Boiling point: 4603°C (8317°F)
Density: 13.31 g/cm³

Common isotopes: Hf-180, Hf-178, Hf-177
State (at STP): Solid
Colour: Silver

# Tantalum

In Greek mythology, Tantalus was the father of Niobe, punished by the gods for sacrificing his son and serving him up at a banquet. This taint of slaughter has spilled over to the eponymous chemical element: tantalum is a metal whose cost is not measured in money, but in blood. The principal tantalum ore, tantalite, is found intermingled with the main niobium ore, columbite, in 'coltan' orebodies. Found in the Democratic Republic of Congo, Rwanda and Venezuela, coltan is considered a 'conflict mineral', associated with illegal smuggling, military conflict, corruption and child slavery.

The reason for its desirability lies in modern technology: every mobile phone contains about 40 mg of tantalum in tiny 'pinhead' capacitors. Each is a miniaturized sandwich of conducting powdered tantalum, separated by a thin layer of insulating tantalum oxide. Like aluminium, titanium and zirconium, tantalum's oxide layer is exceptionally resistant to corrosion. It is one of the few metal elements that can be used for artificial implants in the body, such as metal pins and replacement joints.

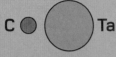

C ● ○ Ta

Atomic radius: 200pm

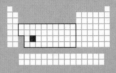

Group 5, Period 6
Transition metal

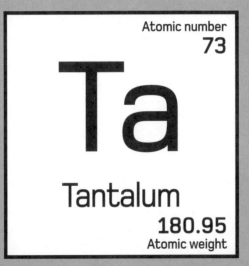

Atomic number
73

# Ta

Tantalum
180.95
Atomic weight

Melting point: 3017°C (5463°F)
Boiling point: 5458°C (9856°F)
Density: 16.654 g/cm³

Common isotopes: Ta-181
State (at STP): Solid
Colour: Silver

# Tungsten

Tungsten's unusual chemical symbol comes from its traditional name wolfram, which means "wolf cream" in German. Traditionally mined in Bohemia and Saxony, this characterful apellation refers to the great quantities of tin consumed during extraction of the metal. The name tungsten itself comes from Swedish for 'heavy stone'.

Element 74 is as hard as they come: taking its place with the heavyweight elements along the bottom row of the periodic table, this super-dense transition metal is largely impervious to attack by acids or bases, and resistant to oxidation. Tungsten is 70 per cent denser than lead and forms the core of armour-piercing bullets. With the highest melting point of any metal (and low expansion when heated) it is ideal material for the electrodes in xenon arc lamps and the filaments of incandescent light bulbs. Pure tungsten is brittle, but mixed with carbon to make tungsten carbide, or alloyed in cobalt tungsten steel, it makes wear-resistant tools.

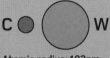
C ◯ W
Atomic radius: 193pm

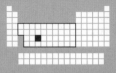
Group 6, Period 6
Transition metal

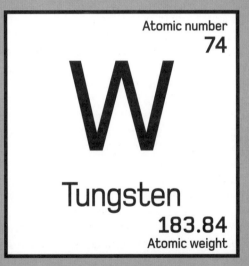

Atomic number
74

# W

## Tungsten

183.84
Atomic weight

Melting point: 3422°C (6192°F)
Boiling point: 5555°C (10,031°F)
Density: 19.25 g/cm³

Common isotopes: W-184, W-186, W-182
State (at STP): Solid
Colour: Silver

# Rhenium

Rhenium was the last stable element to be discovered, and its identification, in 1925 by German chemists Walter Noddack, Ide Tacke and Otto Berg, filled the periodic table's final empty slot. They named it after *Rhenus*, the Latinized name of the German River Rhine. Mostly obtained from the processing of molybdenum ore, Rhenium is more dense than gold and is one of the rarest elements in the Earth's crust. With immense resistance to heat, rhenium is considered a refractory metal. It has the third-highest melting point of all elements, after tungsten and carbon, and the highest boiling point. In the future, it could become more widely used as an industrial catalyst.

Added to tungsten, rhenium makes that tough metal more workable. It is also a crucial ingredient in nickel superalloys – compounds that are used in situations where materials must maintain stiffness at extreme temperatures, such as the turbine blades of jet engines. Incredibly, these blades are grown as a single crystal and are hollow inside to reduce weight.

C 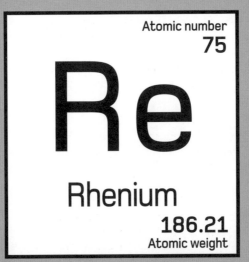 Re

Atomic radius: 188pm

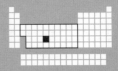

Group 7, Period 6
Transition metal

Atomic number
75

# Re

Rhenium

186.21
Atomic weight

Melting point: 3186°C (5767°F)
Boiling point: 5596°C (10,105°F)
Density: 21.02 g/cm³

Common isotopes: Re-187, Re-185
State (at STP): Solid
Colour: Silver

# Osmium

Osmium is principally known for being the densest chemical element: theoretical values hand the title to its neighbour iridium, but measured density puts osmium first. This shiny silver platinum group metal with a distinct bluish tint is twice as dense as a block of lead of the same volume. Osmium is found in a naturally occurring alloy with iridium, known as osmiridium.

Extremely hard, incompressible and brittle, osmium is almost impossible to beat, bend or form, and completely unworkable. Instead, it is another element whose principal use seems confined to the luxury pen market. Since any hard metal would do the job equally well, the cachet of having a fountain-pen nib made of osmiridium is presumably the whole point, rather than any unique property of osmium. The metal is also notable for the extreme toxicity of its oxide. 'Smoky' osmium tetroxide (the element gets its name from the Greek *osme* meaning 'smell') attacks eyes, lacerates lungs and slips easily into the body through the skin.

C ● **Os**

Atomic radius: 185pm

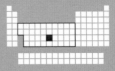

Group 8, Period 6
Transition metal

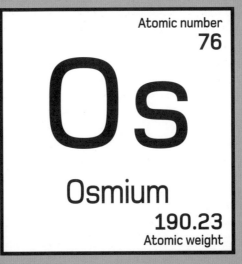

Atomic number
76

# Os

Osmium

190.23
Atomic weight

Melting point: 3033°C (5491°F)
Boiling point: 5012°C (9054°F)
Density: 22.61 g/cm³

Common isotopes: Os-192, Os-190, Os-189
State (at STP): Solid
Colour: Silver-blue

# Iridium

Iridium is a bright platinum group metal with a yellowish tinge. It is one of the rarest elements in the crust – gold is 40 times more abundant Along with its 'sister element' osmium, it is isolated from platinum ores and naturally occurring osmiridium alloy. Officially, it is the second-densest element after osmium, though there's not much in it: a tennis-ball-sized lump of the metal would weigh almost 3 kilograms (6.6 lb). Like osmium, iridium is so hard as to be almost impossible to machine.

A thin band of iridium-enriched muddy sediment in 66-million-year-old clay is the key to explaining the death of the dinosaurs. The 'iridium anomaly' exposed at a roadcut in Gubbio, Italy, contains some 30 times more iridium than normal crust rocks. In 1980, American scientists Walter and Luis Alvarez proposed that the iridium spike was consistent with a 10-kilometre (6-mile) asteroid slamming into the Earth, vapourizing and spreading its heavy metal cargo across the globe. This extinction-level event opened the way for the rise of mammals, and ultimately our own species.

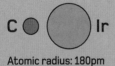

C Ir

Atomic radius: 180pm

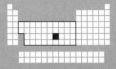

Group 9, Period 6
Transition metal

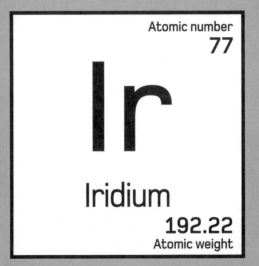

Atomic number
77

**Ir**

Iridium

192.22
Atomic weight

Melting point: 2446°C (4435°F)
Boiling point: 4428°C (8002°F)
Density: 22.56 g/cm³

Common isotopes: Ir-193, Ir-191
State (at STP): Solid
Colour: Silvery yellow

# Platinum

Rarer and more expensive than gold, platinum is the hallmark of prestige and opulence. Discovered by indigenous peoples in South America over 2,000 years ago, it first came to the notice of Europeans in the mid-1500s. Spanish colonists named it *platina*, meaning 'little silver', on account of its appearance, but regarded it as a nuisance impurity, since its high melting point made it hard to separate. Platinum was eventually isolated in 1735 by Antonio de Ulloa.

This beautifully white, dense metal resists corrosion and remains hard when heated – properties that make it a desirable precious metal for use in jewellery, and also a valuable industrial catalyst. Along with palladium and rhodium, platinum is a component of three-way catalytic converters that reduce automobile emissions. These devices convert carbon monoxide to carbon dioxide, nitrogen oxides into nitrogen and oxygen, and split incompletely burned hydrocarbons into carbon dioxide and water.

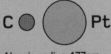

C ○ ⬤ Pt

Atomic radius: 177pm

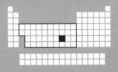

Group 10, Period 6
Transition metal

Atomic number
78

# Pt

Platinum

195.08
Atomic weight

Melting point: 1768.3°C (3214.9°F)
Boiling point: 3825°C (6917°F)
Density: 21.46 g/cm³

Common isotopes: Pt-195, Pt-194, Pt-196
State (at STP): Solid
Colour: White

# Gold

While gold is neither the rarest nor the most expensive element, its price is far less volatile than other noble metals, reflecting its cultural role as a stable, low-risk store for wealth. Used as tender for millennia, gold keeps its value even when – or *especially* when – currencies collapse. Chemically, too, gold is a steady ship: fantastically unreactive, it is untouched by air, water or time and keeps its lustre eternally.

Sheen and scarcity together make gold precious: the 165,446 tonnes mined in human history are barely enough to make a 20-metre (66-ft) cube. Unlike its near neighbours on the periodic table, gold is soft and easy to work – a fingernail-sized scrap can be hammered flat into a square metre (10.75 sq ft) of gold leaf, so thin that light shines through it. 1 gram (0.03 oz) of gold can also be stretched into a 24 kilometre (15-mile) thread. Silver and copper are better conductors of electricity and heat, but since tarnish doesn't dampen gold's abilities, it is used in electronic circuits and spacecraft heat shields.

C 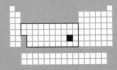 Au

Atomic radius: 174pm

Group 11, Period 6
Transition metal

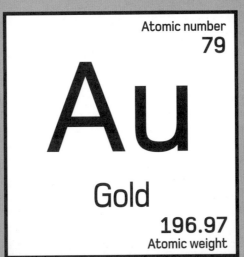

Atomic number
79

Au

Gold

196.97
Atomic weight

Melting point: 1064.2°C (1947.5°F)
Boiling point: 2856°C (5173°F)
Density: 19.282 g/cm³

Common isotopes: Au–197
State (at STP): Solid
Colour: Gold

# Mercury

**M**ercury is an ancient element that is slipping out of sight. Traditionally known as quicksilver, it is one of only two liquid elements on the periodic table under standard conditions, and the only liquid metal. It is easily separated from its main sulfide ore, cinnabar. It was once widelspread – in thermometers, taken as an anti-depressant and a laxative, used as an antiseptic, in paint pigments, batteries and dental fillings – but its high toxicity and tendency to build up in the food chain has led to strict regulation.

With its unique properties, mercury held a fascination for ancient alchemists, who found it a handy reagent, capable of incorporating most other metals into 'amalgams'. Because of its high surface tension, mercury is slippery, but not wet. With little interaction with other surfaces, it trickles, tumbles and clumps into tight balls, but since it doesn't spread, it doesn't feel wet to touch. Highly reflective, mercury can be used to create liquid mirrors. Compact fluorescent light bulbs, fluorescent lamps, and certain car headlamps, meanwhile, use mercury vapour in a sealed tube.

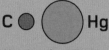

C ● ⬤ Hg

Atomic radius: 171pm

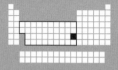

Group 12, Period 6
Transition metal

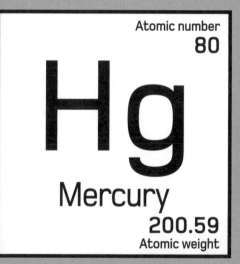

Atomic number
80

Hg

Mercury
200.59
Atomic weight

Melting point: –38.7°C (–37.7°F)
Boiling point: 357°C (675°F)
Density: 13.5336 g/cm³

Common isotopes: Hg–202, Hg–200, Hg–199
State (at STP): Liquid
Colour: Silver

# Thallium

G roup 13's heaviest stable member is another chemical element named after its predominant spectral colours. It was discovered independently by French chemist Claude-Auguste Lamy (1820–78) and English physicist William Crookes (1832–1919) in 1861, using the newly developed technique of spectroscopy. Once again, the new element was the subject of an Anglo-French spat over priority, but Crookes's name for it (from the Greek *thallos*, meaning 'green shoots') won out.

Although thallium is not rare, it is recovered only as a by-product from lead, zinc and copper ore processing. This post-transition metal is known as 'the poisoner's poison', featured by crime writer Agatha Christie in *The Pale Horse*. Soluble thallium salts are highly toxic, and – crucially, if murder is your game – tasteless and odourless when dissolved in water. The body mistakes thallium for potassium, causing a number of slow-acting and painful effects, including nerve damage, but the delayed onset of symptoms makes it hard to trace the source of the illness.

C  Tl

Atomic radius: 156pm

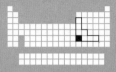

Group 13, Period 6
Post-transition metal

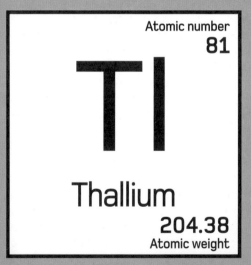

Atomic number
81

Tl

Thallium

204.38
Atomic weight

Melting point: 304°C (579°F)
Boiling point: 1473°C (2683°F)
Density: 11.85 g/cm³

Common isotopes: Tl-205, Tl-203
State (at STP): Solid
Colour: Silver

# Lead

While iron may have been the element that built the Roman Empire, lead may have brought it to its knees. One theory holds that lead added as a 'sweetener' to wine poisoned the emperors, driving their ever more demented behaviour. Lead's chemical symbol comes from the Latin name for lead – *plumbum* – and will be forever associated with plumbing.

Lead is a bluish-white metal that dulls quickly; it is workable and its low melting point makes it easy to cast. As a widely used building material, its softness makes it an effective sealant: spikes in the price of lead are inevitably followed by a spate of thefts of roof flashing. Lead also provides heft to ballistics, and is dense enough to stop gamma and X-rays. For such a heavy metal, lead is also more abundant than might be expected, thanks to the decay chains of radioactive uranium and thorium, both of which end at this stable element. However, it is a notorious neurotoxin, linked to learning disabilities in children: its ubiquity in the environment makes it a menace.

Atomic radius: 154pm

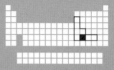
Group 14, Period 6
Post-transition metal

Atomic number
82

Pb

Lead

207.2
Atomic weight

Melting point: 327.5°C (621.4°F)
Boiling point: 1749°C (3180°F)
Density: 11.342 g/cm³

Common isotopes: Pb-208, Pb-206, Pb-207
State (at STP): Solid
Colour: Slate grey

# Bismuth

Early miners gave bismuth the name *tectum argenti*, reflecting their belief that the native mineral was silver in the process of formation. First identified in mines in Germany, bismuth is a dense element, often credited with being the heaviest stable element. However, in 2003, scientists found that its only stable isotope, Bi-209, is in fact very slightly radioactive. Whether a half-life greater than the age of the universe constitutes an 'unstable' isotope is a moot point – technically, Pb-208 is the heaviest truly stable isotope.

Bismuth is brittle and silvery-white, but takes on a rainbow-coloured, iridescent oxide tarnish. It often grows as 'hopper crystals' – spiralling stair-step structures that arise when a crystal grows quicker on its outer edges. Surprisingly for such a heavy metal (and given its proximity to lead and antimony), bismuth is not particularly toxic: it is used in 'pink bismuth' preparations for stomach upset, and also substitutes for toxic lead in lead-free solder, ammunition and 'dragon's egg' fireworks.

C ◯ ⬤ Bi

Atomic radius: 143pm

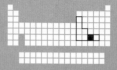

Group 15, Period 6
Post-transition metal

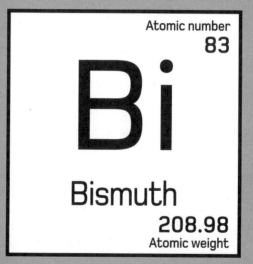

Atomic number
83

# Bi

Bismuth

208.98
Atomic weight

Melting point: 271.6°C (520.9°F)
Boiling point: 1564°C (2847°F)
Density: 9.807 g/cm³

Common isotopes: Bi-209
State (at STP): Solid
Colour: Silver white

# Polonium

Until the death by poisoning of former KGB agent Alexander Litvinenko in London in 2006, polonium was synonymous with Marie Curie. She and her husband Pierre isolated it from uranium ore in 1898, naming it after her native Poland. Their daughter, Irène Joliot-Curie, who developed leukaemia after working with polonium, may well have been the first victim of its radioactivity.

Unlike the other group 16 elements, polonium is a metal. All of its isotopes are radioactive, and as a result, it is very rare: it takes a processed tonne of pitchblende to yield just a milligram of polonium, so most is produced artificially by irradiating Bi-209 with neutrons to form Bi-210, which then decays by beta radiation to Po-210. Only about 100 grams (3.5 oz) are produced each year, almost all in Russian nuclear reactors, so the discovery of this super-rare substance in Litvinenko's system was widely construed as a smoking gun. Polonium is an intense alpha emitter, used as an atomic heat source in thermoelectric generators, in industrial antistatic brushes and, it seems, as an effective killer.

C ◯ Po

Atomic radius: 135pm

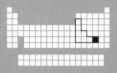

Group 16, Period 6
Post-transition metal

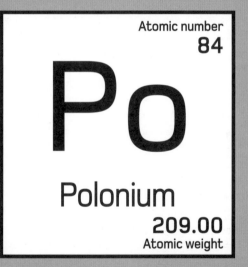

Atomic number
84

# Po

Polonium

209.00
Atomic weight

Melting point: 254°C (489°F)
Boiling point: 962°C (1764°F)
Density: 9.32 g/cm³

Common isotopes: None stable
State (at STP): Solid
Colour: Silver

# Astatine

In theory, all elements up to atomic number 94 occur naturally. However, certain elements, such as astatine and francium, are so unstable as to be almost non-existent in the universe. All primordial stocks of astatine disappeared long ago, while new atoms (forming from the radioactive decay of other elements) break down almost immediately. Of astatine's 39 isotopes, the longest-lived (As-210) has a half-life of 8.1 hours, and many have half-lives measured in nanoseconds. So the element's name, derived from Greek *astatos*, meaning 'unstable', is well chosen.

Astatine is officially the rarest element. At any one time there is an estimated 30 grams (1 oz) in the Earth's crust. As a result, astatine has a total lack of applications, although there may be some potential in radiotherapy. It is considered the heaviest of the halogens (since ununseptium's chemistry is unknown), but its properties are theoretical. If enough could be collected astatine would be solid, but would vaporize instantly due to the heat generated by its intense radiation.

C  At

Atomic radius: Unmeasured

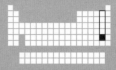

Group 17, Period 6
Halogen

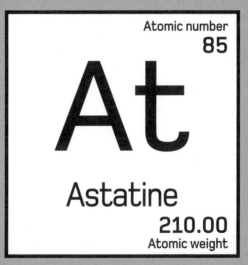

Atomic number
85

At

Astatine
210.00
Atomic weight

Melting point: 302°C (576°F)
Boiling point: 337°C (639°F)
Density: 7 g/cm³

Common isotopes: None stable
State (at STP): Solid
Colour: Silver

# Radon

Element 86 is the silent killer in the basement. This colourless, odourless gas, produced by the radioactive decay of uranium, is barely detectable and radioactive itself. It builds up in poorly ventilated ground floors, cellars and mines in areas with uranium-containing bedrocks such as granite and shale. You really don't want to breathe in radon: while the alpha particles it emits are blocked by skin, they wreak havoc in the lungs. What's more, its decay products – the so-called radon daughters – are a cocktail of of polonium, lead and bismuth radioisotopes. They leave a legacy of damage that makes radon inhalation the second most common cause of lung cancer after smoking.

Radon is the heaviest noble gas. It was discovered in 'emanations' from radioactive radium, actinium and thorium. Although it was first isolated in 1910 by William Ramsay (who favoured the name niton), credit for the discovery goes to German physicist Friedrich Ernst Dorn (1848–1916), who in 1900 repeated Ernest Rutherford's earlier experiments on thorium (see page 348).

C 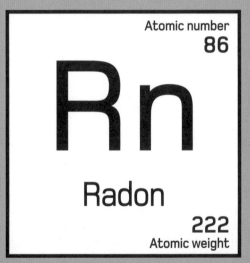 Rn
Atomic radius: 120pm

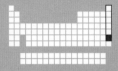

Atomic number
**86**

# Rn

## Radon

222
Atomic weight

Melting point: −71°C (−96°F)
Boiling point: −61.9°C (−79.4°F)
Density: 0.00973 g/cm³

Common isotopes: None stable
State (at STP): Gas
Colour: Colourless

# Francium

One of the two elements on the periodic table named after France (the other is gallium – see page 230), francium is an exceedingly rare, radioactive alkali metal. Discovered in 1939 by Marie Curie's student Marguerite Perey (1909–75), it sits at the bottom of group 1 of the periodic table.

No weighable sample of francium has ever existed: this element's most stable isotope has a half-life of 22 minutes. Even so, one might be bold enough to make certain predictions about the bulk metal. Since melting points decrease down group 1, francium would likely have a lower melting point than caesium. Its radioactivity produces a lot of energy,  so it would be warm to touch, and may even be a liquid at room temperature. Francium could also be a coloured metal, like caesium. Despite the general trend of increasing reactivity down the elements of group 1, chemists think that it would be slightly harder to remove francium's outermost electron, making it less reactive than caesium.

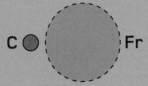

C ◯ Fr

Atomic radius: Unmeasured

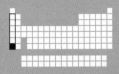

Group 1, Period 7
Alkali metal

Atomic number
87

# Fr

## Francium

223
Atomic weight

Melting point: 26.9°C (80.3°F)
Boiling point: 677°C (1251°F)
Density: 1.87 g/cm³

Common isotopes: None stable
State (at STP): Solid
Colour: Silver

# Radium

The radioactive elements that immortalized the names of Pierre Curie (1859–1906) and Marie Curie (1867–1934) also caused their premature deaths. Among them, radium is the most famous and most dangerous. The husband and wife team discovered element 88 in the uranium ore pitchblende in 1898 and named it radium for its powerfully emitted rays. With a half-life of 1,600 years, radium-226 occurs naturally at about a few milligrams per tonne of ore.

Radium is the archetypal radioactive material, spitting out prodigious amounts of alpha, beta and gamma radiation – enough to make a lump of radium glow green and feel warm to the touch. Early on, it was used as a curative: spas advertised its recuperative powers, and it was included in toothpastes and medicines. It was also used in radiotherapy treatment of cancer. Self-illuminating paints used radiation from radium compounds to excite fluorescent chemicals. However, such materials were later found to be highly carcinogenic.

C ● ··· Ra

Atomic radius: Unmeasured

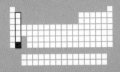

Group 2, Period 7
Alkaline earth metal

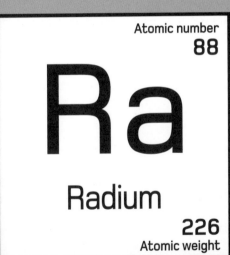

Atomic number
88

# Ra

Radium

226
Atomic weight

Melting point: 700°C (1292°F)
Boiling point: 1737°C (3159°F)
Density: 5.5 g/cm³

Common isotopes: None stable
State (at STP): Solid
Colour: Silver-white

# Actinium

Element 89 gives its name to the sweep of elements that underscore the periodic table like the spacebar of an old-fashioned typewriter. This 'actinoid' series of ephemeral and short-lived elements (see page 152) marks the official starting point of nuclear chemistry – however, just as the lanthanoids include scandium and yttrium as rare earth metals, so the actinoids appropriate radium and radon. None of them have stable isotopes, and only the first four occur on Earth naturally.

André-Louis Debierne (1874–1949), a friend of Pierre and Marie Curie, discovered actinium in 1899. He isolated it from residues of that melange mineral, pitchblende, which also yielded uranium, polonium, radon, thorium and protactinium. Glowing blue, actinium is visibly radioactive. But while it occurs naturally, it is more commonly produced by bombarding radium-226 with neutrons in nuclear reactors. Actinium-227 is used in research as a neutron source, while Ac-225 may play a role in 'magic bullet' cancer treatments that selectively target tumour cells.

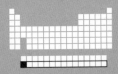

C ◯ ● Ac

Atomic radius: Unmeasured

Period 7
Actinoid

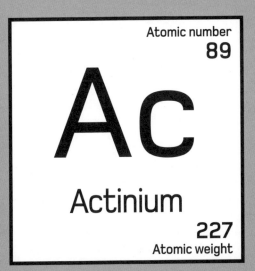

Atomic number
89

Ac

Actinium

227
Atomic weight

Melting point: 1050°C (1922°F)
Boiling point: 3198°C (5788°F)
Density: 10.07 g/cm³

Common isotopes: None stable
State (at STP): Solid
Colour: Silver

# Thorium

O n 16 June 1994, FBI officers swooped on a house in a suburb of Detroit. Their target was 17-year-old David Hahn who had built a breeder reactor in his garden shed and was attempting to produce radioactive actinoid elements – all as a project for his Eagle Scout badge. Hahn had collected his thorium fuel from camping lanterns: thorium oxide has the highest melting point of all oxides and has been used for incandescent gas mantles since 1891 (alpha particles emanating from the mantles are stopped easily by the lantern glass).

In fact, reactors like Hahn's are touted as the answer to nuclear power's many problems. Firstly, thorium may be three times more abundant in the crust than uranium and is widely distributed in ores such as monazite. Secondly, thorium reactors are inherently safer than water reactors: they don't require a vast concrete containment building around the reactor, and their waste is safe within ten years. Finally, the amount of plutonium they produce is limited, reducing the proliferation of nuclear weapons.

Atomic radius: Unmeasured

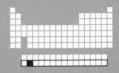
Period 7
Actinoid

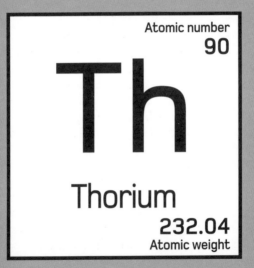

Atomic number
90

Th

Thorium

232.04
Atomic weight

Melting point: 1842°C (3348°F)
Boiling point: 4788°C (8650°F)
Density: 11.72 g/cm³

Common isotopes: Th-232
State (at STP): Solid
Colour: Silver

# Protactinium

Element 91 is another of Mendeleev's predictions. In 1871 the Russian chemist spotted a 'hole' in the periodic table and predicted an element with atomic weight intermediate between thorium and uranium would eventually fill it. It was eventually found in 1918 by Lise Meitner (1878–1968) and Otto Hahn but, perplexingly, the element weighed less than thorium. This might have caused consternation had the discovery been made before Henry Moseley measured atomic number (see page 76), but since atomic number, not weight, is a truer organizing principle for the periodic table, this rare 'pair reversal' is merely an interesting quirk of the elements.

'Proto-actinium', as it was first called, was so named because its decay through the loss of an alpha particle – two protons and two neutrons – produces actinium. Itself a decay product of uranium, the element occurs naturally. The largest deposit, however, is the 125 grams (4.4 oz) artificially produced in 1961 by the UK's Atomic Energy Authority using 60 tonnes of radioactive waste.

C 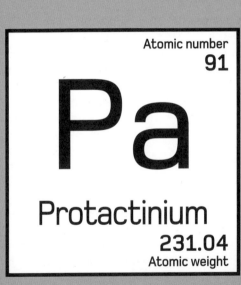 Pa

Atomic radius: Unmeasured

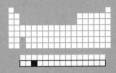

Period 7
Actinoid

Atomic number
91

# Pa

Protactinium

231.04
Atomic weight

Melting point: 1568°C (2854°F)
Boiling point: 4027°C (7281°F)
Density: 15.37 g/cm³

Common isotopes: None stable
State (at STP): Solid
Colour: Silver

# Uranium

Uranium is the element that ushered in the atomic age. Discovered in 1798 by German chemist Martin Klaproth, it was named after the recently discovered planet Uranus. Until 1938, it was a novelty additive to glass that made shimmery glassware but, in 1938, Otto Hahn and Fritz Strassmann (1902–80) split uranium's atomic nucleus. Hungarian physicist Leó Szilárd (1898–1964) figured out how a chain reaction could drive self-sustaining nuclear fission to yield colossal amounts of energy, or destructive power. And in 1942, the world's first nuclear reactor – literally a pile of fissile uranium – went critical in a squash court at the University of Chicago. Complete fission of 1 kilogram (2.2 lb) of U-235 produces as much energy as 1,500 tonnes of coal, but natural uranium ore contains a mere 0.7204 per cent of fissile U-235, so it must be 'enriched' to between 3 and 5 per cent before use. Depleted uranium – U-238 with almost all the U-235 removed – contains hardly any fissile material. It is reactive, pyrophoric and poisonous. Hugely dense, it is used both for vehicle armour and for armour-piercing rounds.

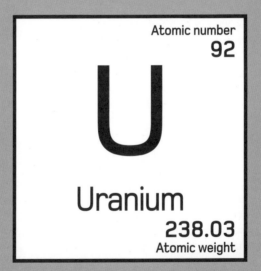

C ⬤ U
Atomic radius: Unmeasured

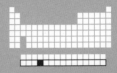

Period 7
Actinoid

Atomic number
92

U

Uranium
238.03
Atomic weight

Melting point: 1132.2°C (2070°F)
Boiling point: 4131°C (7468°F)
Density: 18.95 g/cm³

Common isotopes: U-238
State (at STP): Solid
Colour: Silver

# Neptunium

In 1940, American physicists Edwin McMillan (1907–91) and Philip Abelson (1913–2004) created the very first element heavier than uranium at the University of California, Berkeley. Using a technique pioneered by Enrico Fermi (1901–54), they fired 'slow' (low-energy) neutrons at a thin plate of uranium-238 oxide: after many impacts, some neutrons would stick, creating a U-239 nucleus that would then undergo beta decay, leaving it with an extra proton and heavier mass. Naturally, this new element was named after the next planet beyond Uranus.

This landmark discovery opened the door to the upper reaches of the periodic table. No 'transuranic' elements exist naturally, except as a by-product of a minor decay chain of uranium-235. McMillan and Abelson produced Np-239, which has a half-life of just two and a half days, but Np-237 has a half-life of 2 million years. It is often recovered from nuclear waste, but since it is fissionable with a critical mass of just 60 kilograms (132 lb), its use is strictly monitored.

C ◯ ⬤ Np

Atomic radius: Unmeasured

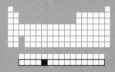

Period 7
Actinoid

Atomic number
93

Np

Neptunium

237
Atomic weight

Melting point: 644°C (1191°F)
Boiling point: 4000°C (7232°F)
Density: 20.45 g/cm³

Common isotopes: None stable
State (at STP): Solid
Colour: Silver

# Plutonium

Appropriately, when the NASA spaceprobe New Horizons blazed past the dwarf planet Pluto in 2015, it carried on board a small amount of plutonium, the last in a trio of elements named after the outer planets. New Horizons is not nuclear powered in the conventional sense, however: its on-board power comes from a radioisotope thermoelectric generator (RTG), which uses the heat from a pellet of Pu-238 to produce electricity in the cold depths of the outer solar system.

Plutonium was first synthesized in 1940 as part of the USA's secret wartime Manhattan Project, but its discovery remained undisclosed until 1948. By then, however, its arrival had already been announced in terrible fashion with the 'Fat Man' plutonium-core bomb detonated over Nagasaki, Japan, on 9 August 1945. Metallic plutonium is pyrophoric: heat generated by its alpha decay makes a lump of the metal glow with orange-red heat and slowly destroys it from within. Weapons-grade plutonium is produced in breeder reactors and is strictly monitored.

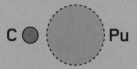

C ◯ ◯ Pu

Atomic radius: Unmeasured

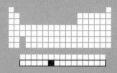

Period 7
Actinoid

Atomic number
94

# Pu

Plutonium

244
Atomic weight

Melting point: 639.4°C (1182.9°F)
Boiling point: 3228°C (5842°F)
Density: 19.84 g/cm³

Common isotopes: None stable
State (at STP): Solid
Colour: Silver

# Americium

L ike plutonium before it, americium was produced by
Glenn T. Seaborg and a group of Berkeley scientists
working at the University of Chicago as part of the top
secret Manhattan Project (see page 356). Evidently, some
plutonium-239 captured two neutrons to form Pu-241, which
then released a beta particle to transform into Am-241.
Americium was isolated in 1944, but wartime secrecy prevented
Seaborg from publishing – he eventually spilled the beans in
1945 on a children's radio show.

Am-241, the most common isotope, is a stronger emitter of alpha
particles than radium and is used in smoke detectors. Charged
particles coming off an americium dot allow electric current to
flow across an air gap but, if smoke intervenes, the alpha particles
are blocked and the drop in current sets off an alarm. Americium
is the only widely available actinoid, but fortunately, rogue bomb-
builders would need to marshal 180 billion smoke detectors before
they had enough fissile material for a critical mass.

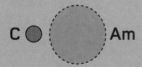
Atomic radius: Unmeasured

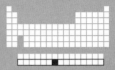
Period 7
Actinoid

Atomic number
95

# Am

Americium

243
Atomic weight

Melting point: 1176°C (2149°F)
Boiling point: 2607°C (4725°F)
Density: 13.69 g/cm³

Common isotopes: None stable
State (at STP): Solid
Colour: Silver

# Curium

The first two transuranic elements to be synthesized were
made by bombarding heavy elements with slow neutrons.
However, the efficacy of neutron bombardment dwindled sharply
after americium. Neutrons just wouldn't cut it and, instead,
Glenn T. Seaborg and his group created curium in 1944 by
bombarding plutonium with heavier alpha particles.

Curium is named in honour of Pierre and Marie Curie, discoverers
of the first radioactive elements, and follows the chemical
precedents set by its corresponding f-block element, gadolinium.
Like all actinoids, curium is a silver-white, dense metal, and is a
starting material for synthesizing heavier transuranic elements.
Most actinoids are highly toxic: since none were around during
the period in which life evolved on Earth, they play no part in body
chemistry. Curium is also among the most strongly radioactive
of the isolatable elements – so much so that it glows purple. Tiny
lumps are used as alpha emitters in instruments aboard the
Mars Rovers and Philae comet lander.

C ● ⬤ Cm

Atomic radius: Unmeasured

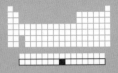
Period 7
Actinoid

Atomic number
96

# Cm

Curium

247
Atomic weight

Melting point: 1340°C (2444°F)
Boiling point: 3110°C (5630°F)
Density: 13.51 g/cm³

Common isotopes: None stable
State (at STP): Solid
Colour: Silver-white

# Berkelium

By the late 1940s, the dreams of the early alchemists had not only been realized, but also turned into an industry. At the 1.5-metre (60-in) cyclotron at Berkeley, California (see page 114), Glenn T. Seaborg and his team, including Albert Ghiorso and Stanley Thompson (1912–76), were routinely transmuting elements. Between 1949 and 1950, they forged two more elements (97 and 98) never seen before on Earth. Naming followed the method established for elements 95 and 96, which mirror their corresponding lanthanoid: so just as terbium (element 65) took its name from Ytterby (see page 298), Seaborg and co-workers named element 97 after its birthplace.

Early transactinoid elements are stable enough to allow their chemistry to be studied: berkelium's longest-lived isotope, Bk-247, has a half-life of 1,380 years. Like other actinoids, it easily enters aqueous solutions, forming compounds with a 3+ or 4+ oxidation state. Berkelium's main use, however, is as a target for synthesizing other heavy elements.

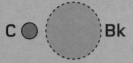

C ● ⬤ Bk

Atomic radius: Unmeasured

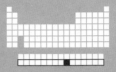

Period 7
Actinoid

Atomic number
97

# Bk

Berkelium

247
Atomic weight

Melting point: 986°C (1807°F)
Boiling point: 2627°C (4761°F)
Density: 14.79 g/cm³

Common isotopes: None stable
State (at STP): Solid
Colour: Silver

# Californium

A 2014 *Business Insider* report listed californium as the world's most expensive metal: Cf-252 costs approximately US$27 million per gram (compared to a gold price of around US$40/g). However, since the most common isotope has a half-life of less than three years, californium is not a particularly attractive investment prospect.

The element was synthesized by Glenn T. Seaborg's team at the Berkeley 'Rad Lab' in 1950, by bombarding curium with alpha particles. Due to its radioactivity, some californium compounds have the beautiful and uncanny property of self-luminescence. It is the heaviest element on the periodic table that is stable enough for its chemistry to be practically studied. Unusually for a transuranic element, californium has several applications. This is because Cf-252 is a neutron emitter: just one microgram sprays out around 139 million neutrons per minute. It is therefore useful to kick-start nuclear reactors, and its penetrating neutrons are used in detection instruments and certain cancer treatments.

C  Cf

Atomic radius: Unmeasured

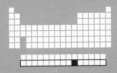

Period 7
Actinoid

Atomic number
98

# Cf

Californium

251
Atomic weight

Melting point: 900°C (1652°F)
Boiling point: 1470°C (2678°F)
Density: 15.1 g/cm³

Common isotopes: None stable
State (at STP): Solid
Colour: Silver

# Einsteinium

The dawn of the atomic age introduced an entirely new vocabulary: the products of the Manhattan Project brought a chilling familiarity to neologisms such as fallout, ground zero and mushroom cloud. And fallout from the testing of the first hydrogen bomb, codenamed 'Ivy Mike', in 1952 also yielded two actinoid elements. The lighter of these two new elements was named after Albert Einstein – an odd choice, since he was not a nuclear physicist. Although as author of the equation $E=mc^2$, Einstein could be considered the father of nuclear weapons, it was certainly not a legacy he was happy with, and he campaigned ardently against their use.

Today, einsteinium is mostly created by bombardment of plutonium with neutrons. The element is a typical late actinoid – a silvery metal that glows in the dark as radiation breaks down its internal bonds and releases energy. Einsteinium's most common isotope, Es-253, has a half-life 20.47 days; that of its longest-lived isotope is 472 days.

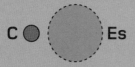

C ● ⟨ ⟩ Es

Atomic radius: Unmeasured

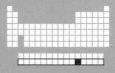

Period 7
Actinoid

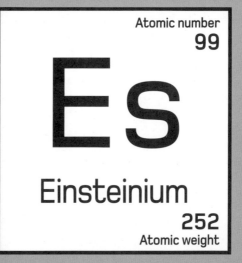

Atomic number
99

# Es

Einsteinium

252
Atomic weight

Melting point: 860°C (1580°F)
Boiling point: 996°C (1824.8°F)
Density: 8.84 g/cm³

Common isotopes: None stable
State (at STP): Solid
Colour: Silver

# Fermium

Element 100, like einsteinium, was another new element sifted from the ashes of the first H-bomb test. Fermium was discovered among hundreds of tons of radioactive ashes and irradiated coral from the ex-island of Elugelab in the Pacific Enewetak Atoll. Pounded by intense neutrons during the explosion, uranium in the initiator device had transmutated into a shower of heavier elements, including roughly 200 atoms of element 100. Researchers at Berkeley named the element fermium after Enrico Fermi, a pioneer of nuclear physics.

The discoveries of einsteinium and fermium were kept secret for three years, but were rapidly declassified in 1955 on news of a competing claim. No solid sample of fermium has ever been produced. Its longest-lived isotope, Fm-257, has a half-life of 100 days. As an alpha emitter it is a potentially useful medical radioisotope, but it is frustratingly elusive: Fm-257 produced in nuclear reactors readily accepts another neutron to become the wildly unstable Fm-258, which vanishes in milliseconds.

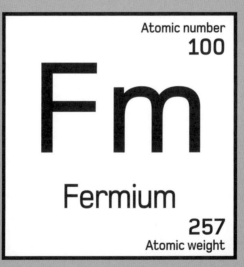

C ◯ ⬤ Fm

Atomic radius: Unmeasured

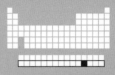

Period 7
Actinoid

Atomic number
100

# Fm

Fermium

257
Atomic weight

Melting point: 852°C (1566°F)
Boiling point: Unknown
Density: Unknown

Common isotopes: None stable
State (at STP): Unconfirmed
Colour: Unknown

# Mendelevium

B y the mid-1950s, a new trick was needed to keep pushing the boundaries of stable matter. Elements heavier than fermium – the so-called transfermium elements – are decidedly odd, and their short-lived isotopes are produced in mere handfuls of atoms.

All elements heavier than uranium are non-primordial. With no stable isotopes and half-lives much shorter than the age of the Earth, all of their 'original stocks' have long since decayed. This is why the superheavy elements had to be synthesized – they couldn't be discovered. Fermium marked a limit to the elements that could be synthesized by neutron capture in a nuclear reactor: element 101, named in honour of Dmitri Mendeleev, was prepared at Berkeley by bombarding einsteinium with helium ions in the 1.5-metre (60-in) cyclotron. The resulting 17 Md-256 atoms had a half-life of just 87 minutes. Although such synthetic elements are created artificially on Earth, they are 'natural', in the sense that they can still be produced in supernovae.

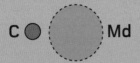

C ⬤ ⬤ Md

Atomic radius: Unmeasured

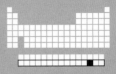

Period 7
Actinoid

Atomic number
101

Md

Mendelevium

258
Atomic weight

Melting point: 827°C (1521°F)
Boiling point: Unknown
Density: Unknown

Common isotopes: None stable
State (at STP): Solid
Colour: Unknown

# Nobelium

Element 102 is among the shorter-lived of the mayfly actinoids. Like all transfermium elements, nobelium can only be produced in a particle accelerator: its most stable isotope, No-259, has a half-life of one hour, and the No-255 isotope most commonly produced has a half-life of just three minutes. When an element's life is this brief, its hard to do much chemistry with it.

Nobelium's discovery was the first shot in the 'transfermium wars', a controversy that dogged the naming of elements 104 to 109. A Swedish group advanced the first claim in 1957, suggesting the name nobelium, after industrialist and Nobel Prize benefactor Alfred Nobel. The name was accepted with undue haste. At Berkeley, they fired up their new Heavy Ion Linear Accelerator (HILAC), peppering curium targets with carbon ions, but couldn't replicate the Swedish result. Credit finally went to a Soviet team but, to their chagrin, the name nobelium was retained over their favoured joliotium (honouring Irène Joliot-Curie).

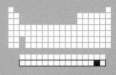

C ● ◯ No

Atomic radius: Unmeasured

Period 7
Actinoid

Atomic number
102

No

Nobelium

259
Atomic weight

Melting point: 827°C (1521°F)
Boiling point: Unknown
Density: Unknown

Common isotopes: None stable
State (at STP): Unconfirmed
Colour: Unknown

# Lawrencium

When academic funnyman Tom Lehrer wrote his famous periodic table song 'The Elements' in 1959 there were only 102 elements. But with discoveries of new elements coming thick and fast, he covered himself with a coda: 'These are the only ones of which the news has come to Harvard, And there may be many others, but they haven't been discovered'.

Sure enough, in 1961 along came element 103. It was produced by Albert Ghiorso and co-workers at the 'Rad Lab' at Berkeley, using their HILAC accelerator to bombard a californium target with boron nuclei. They named the element lawrencium, in honour of American nuclear physicist Ernest Lawrence, the Nobel-laureate director of their laboratory, who had invented the cyclotron and passed away in 1958. Chemically, lawrencium is a heavier lutetium and slots into the periodic table as the final member of the actinoid series, as Glenn T. Seaborg had predicted in 1949. However, just like lutetium, it could equally be counted as a member of group 3.

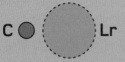

C ◯ ⬤ Lr

Atomic radius: Unmeasured

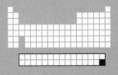

Period 7
Actinoid

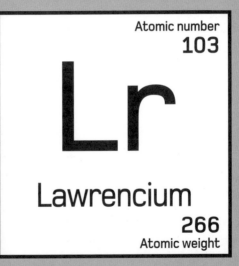

Atomic number
103

# Lr

## Lawrencium

266
Atomic weight

Melting point: 1627°C (2961°F)
Boiling point: Unknown
Density: Unknown

Common isotopes: None stable
State (at STP): Unconfirmed
Colour: Unknown

# Rutherfordium

The final line of the periodic table currently runs from atomic numbers 104 to 118. Positioned in transition metal territory, these transactinoids are d-block elements (see page 122), and the superheavyweights of the chemical world. All of them are violently radioactive, and all of them must be artificially synthesized in particle accelerators.

When a substance is produced in such small quantities it is very hard to say anything definite about it. Individual atoms, or nano-sized pieces, behave in radically different ways to bulk materials. Large collections of atoms develop interactions and crystal structures that affect properties such as ionization energies, melting points, solubility and response to electromagnetic rays in unpredictable ways. Rutherfordium, first synthesized in 1964 and named for New Zealand physicist Ernest Rutherford, is a case in point: its properties are largely theoretical, but it is presumed to be a solid and may even be one of the densest elements, thanks to the lanthanoid contraction (see page 306).

C ◯ Rf

Atomic radius: Unmeasured

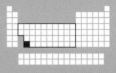

Group 4, Period 7
Transition metal

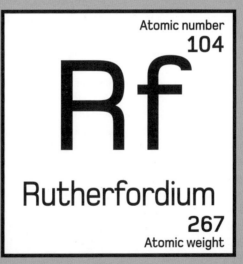

Atomic number
104

Rf

Rutherfordium

267
Atomic weight

Melting point: 2127°C (3861°F)
Boiling point: 5527°C (9981°F)
Density: 23.2 g/cm³

Common isotopes: None stable
State (at STP): Solid
Colour: Unknown

# Dubnium

After enjoying a post-war decade or so of unchallenged supremecy at synthesizing superheavy elements, the USA found that its place at the head of the periodic table was no longer guaranteed. Priority for the discovery of elements 104 to 109 was fought over bitterly, with rival institutions in Berkeley, USA, Dubna, Russia and Darmstadt, Germany, digging their heels in over their own element names. The Americans referred to 104 as rutherfordium; the Soviets called it kurchatovium. Element 105, meanwhile, was known as hahnium by the Americans, while Russia insisted on nielsbohrium. Scientists that have spent decades creating atoms that survive for tiny fractions of a second are not inclined to give in without a fight, and Cold War geopolitics added to the stand-off. In the end, the International Union of Pure and Applied Chemistry (IUPAC) had to appease both sides. To this end, they formally ruled on the transactinoid names in 1997, adopting a new name for element 105. Dubnium is named for the Russian home city of the Joint Institute for Nuclear Research, mirroring the previously named element 97, berkelium.

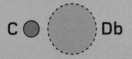
C ⬤ Db

Atomic radius: Unmeasured

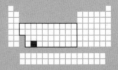

Group 5, Period 7
Transition metal

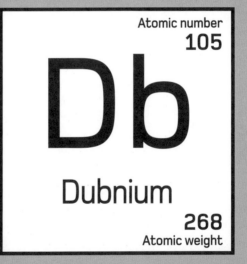

Atomic number
105

# Db

Dubnium

268
Atomic weight

Melting point: Unknown
Boiling point: Unknown
Density: 29.3 g/cm³

Common isotopes: None stable
State (at STP): Solid
Colour: Unknown

# Seaborgium

The chemistry of the superheavy elements was undeniably the domain of Glenn T. Seaborg. Italian physicist Enrico Fermi had the original idea of synthesizing new elements, but it was Seaborg who got the method to work, graduating from neutron irradiation to atom smashing. He also pioneered the chemical techniques used to study the compounds of these short-lived elements, often working with a mere handful of atoms. Realizing that the new elements were homologous to the f-block lanthanoids, rather than d-block transition metals, Seaborg proposed the actinoid series in 1945. In doing so, he rewrote the periodic table – the most significant change since its inception in 1869.

In 1974, there was more controversy over the discovery and naming of element 106, with competing claims from US and Soviet teams. Despite granting the Americans priority, IUPAC refused their name on the grounds that Seaborg was still alive. The American Chemical Society stuck to their guns, and thus seaborgium became the only element named after a living person.

C  Sg

Atomic radius: Unmeasured

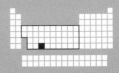

Group 6, Period 7
Transition metal

Atomic number
106

Sg

Seaborgium

269
Atomic weight

Melting point: Unknown
Boiling point: Unknown
Density: 35.0 g/cm³

Common isotopes: None stable
State (at STP): Unconfirmed
Colour: Unknown

# Bohrium

Synthesized in 1981 at the Institute for Heavy Ion Research in Darmstadt, Germany (the Gesellschaft für Schwerionenforschung or GSI), element 107 was named for the Danish physicist Niels Bohr. In 1913, Bohr described a quantum mechanical model of the atom, showing how electrons occupied specific energy levels around the nucleus (see page 74). This paved the way for a valence theory of bonding and simultaneously explained the way that the periods of the periodic table 'turn over'. Bohrium was created by 'cold fusion' of bismuth and chromium ions to create a Bh-262 isotope with a half-life of milliseconds (when ions fuse 'hot', the energy of the collision often breaks them apart again). Bohrium's most stable isotope, Bh-270, has a half-life of 61 seconds. So-called relativistic effects in elements 104 and 105 (caused by electrons moving at significant fractions of the speed of light) threatened to bust the periodic law of elements. The behaviour of seaborgium and bohrium, however, follows their group chemistry, thus restoring faith in the periodic law as an underlying, universal property of matter.

C ◯ ⬤ Bh

Atomic radius: Unmeasured

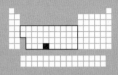

Group 7, Period 7
Transition metal

Atomic number
107

# Bh

## Bohrium

270
Atomic weight

Melting point: Unknown
Boiling point: Unknown
Density: 37.1 g/cm³

Common isotopes: None stable
State (at STP): Unconfirmed
Colour: Unknown

# Hassium

Hassium is one of two elements named after the GSI Institute for Heavy Ion Research in Darmstadt, Germany. The first to attempt synthesis of element 108 was a Russian team at the Joint Institute for Nuclear Research (JINR) in Dubna. But despite claims of success in 1978 and 1983, the honour of the official discovery went to the German team. They chose to name it in reference to *Hassia,* the Latin name of the state of Hesse in Germany where the institute is located.

The team at GSI produced three atoms of Hs-265, with a half-life of a mere 1.95 milliseconds, by firing iron nuclei at a lead target. In 2001, five atoms of the longer-living Hs-269 (1.42-second half-life) were assembled in order to confirm hassium's chemical properties. Forming a volatile tetroxide, it is markedly similar to osmium, though heavier. Density, of course, is a bulk property with no real meaning on the atomic scale, but hassium in bulk is expected to be a dense, silvery white metal.

C ● ⬤ Hs

Atomic radius: Unmeasured

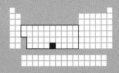

Group 8, Period 7
Transition metal

Atomic number
108

# Hs

Hassium

269
Atomic weight

Melting point: Unknown
Boiling point: Unknown
Density: 40.7 g/cm³

Common isotopes: None stable
State (at STP): Unconfirmed
Colour: Unknown

# Meitnerium

Which is better – winning a Nobel Prize or having an element named after you? These are the sort of imponderables that periodic table buffs love to chew over, and in the case of Lise Meitner the question is particularly germane. As well as co-discovering protactinium with Otto Hahn, the Austrian physicist played a crucial role in discovering nuclear fission in 1938. In 1944, however, the Nobel committee awarded the prize in chemistry for this work to Hahn alone – a blatant case of women's achievements in science being overlooked.

When element 109 was synthesized in Germany in 1982, the team at the GSI chose to honour Meitner. Their suggestion went uncontested – one of the few transactinoid elements not associated with controversy over its naming. Meitnerium was produced by slamming iron nuclei into a bismuth target. The result after billions of collisions was a single atom of Mt-266, with a half-life of 1.7 milliseconds. Meitnerium's most stable isotope Mt-278, meanwhile, has a half-life of 7.6 seconds.

C  Mt

Atomic radius: Unmeasured

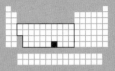

Group 9, Period 7
Transition metal

Atomic number
109

# Mt

Meitnerium

278
Atomic weight

Melting point: Unknown
Boiling point: Unknown
Density: 37.4 g/cm³

Common isotopes: None stable
State (at STP): Unconfirmed
Colour: Unknown

# Darmstadtium

Element 110's name acknowledges the closest city to Germany's GSI nuclear research institute (although since such facilities can't be located safely near large population centres, the GSI is actually based in the outlying suburb of Wixhausen). Hence, following its confirmation, a joke went around that the new transactinoid was originally going to be called wixhausium.

Researchers produced the radioactive element by bombarding a lead target with nickel ions. Darmstadtium is another fleeting element: when it was first created artificially, the single atom winked into being and winked out of existence in just 179 microseconds – barely enough time to register its reality. Darmstadtium-281 is the longest-lived isotope, with a half-life of 11 seconds. Darmstadtium's chemistry has not been experimentally verified, but it is presumed to be a noble metal. Elements 104 to 112 form a fourth series of transition metals. Sitting at the bottom of group 10, darmstadtium should behave in a similar way to nickel, palladium and platinum.

C ⬤ ⦙ ⦙ Ds

Atomic radius: Unmeasured

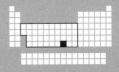

Group 10, Period 7
Transition metal

Atomic number
110

# Ds

## Darmstadtium

281
Atomic weight

Melting point: Unknown
Boiling point: Unknown
Density: 34.8 g/cm³

Common isotopes: None stable
State (at STP): Unconfirmed
Colour: Unknown

# Roentgenium

Element 111 was named after the German physicist Wilhelm Conrad Röntgen (1845–1923), the discoverer of X-rays. Thankfully, by the 1990s, the rancour and discontent that mired each new transactinoid discovery had dissipated, to be replaced by a spirit of international cooperation. IUPAC had also learned some lessons too, emerging chastened and cautious in this new era. Although element 111 was synthesized in 1994, it took another ten years to confirm the discovery and ratify the name.

Roentgenium was discovered at GSI in Darmstadt by a team led by Sigurd Hofmann – another titan of transactinoid chemistry. Once again, the discovery was made from a single atom, produced by smashing nickel nuclei into a bismuth target at high energies. The Rg-272 produced had a half-life of 3.8 milliseconds. The most stable known isotope, Rg-281, has a half-life of 26 seconds. Sitting beneath gold on the periodic table, element 111 is expected to be chemically analogous to the coinage metals, copper, silver and gold.

C 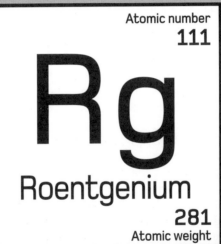 Rg

Atomic radius: Unmeasured

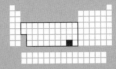

Group 11 Period 7
Transition metal

Atomic number
111

Rg

Roentgenium

281
Atomic weight

Melting point: Unknown
Boiling point: Unknown
Density: 28.7 g/cm³

Common isotopes: None stable
State (at STP): Solid
Colour: Unknown

# Copernicium

Darmstadt scientists first claimed to have created two atoms of element 112 in 1996, in their 120-metre (394-ft) particle accelerator. Firing a high-energy beam of zinc nuclei at a rotating lead target for a week, they produced a lone atom of element 112. Finally in 2000, the GSI team created a second atom. Sigurd Hofmann, leader of the teams that also discovered darmstadtium and roentgenium, described the step-by-step process as a 'difficult and stony path'.

Even more torturous, however, were the 13 years of formal review it took to ratify the discovery. In 2009, IUPAC acknowledged the German team's priority and accepted their name of copernicium. Although Polish astronomer Nicolaus Copernicus might seem a strange choice, honouring a widely acknowledged scientific hero was at least uncontroversial. Chemical experiments confirm that copernicium is a transition element, adding weight to the theory that the periodic table's bottom row forms a fourth series of heavy transition metals.

C 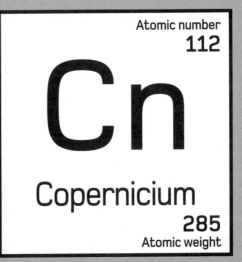 Cn
Atomic radius: Unmeasured

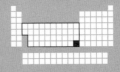

Group 12, Period 7
Transition metal

Atomic number
112

Cn

Copernicium

285
Atomic weight

Melting point: Unknown
Boiling point: 84°C (183°F)
Density: 23.7 g/cm³

Common isotopes: None stable
State (at STP): Unconfirmed
Colour: Unknown

# Ununtrium

Ununtrium is a placeholder name for element 113 ('un-un-trium' quite literally spells out in letters the numbers '1-1-3'), but this will not be its name for much longer. A press release from IUPAC on December 30, 2015 confirmed the discoveries of the final four elements of the periodic table (elements 113, 115, 117 and 118). Formal recognition by IUPAC grants priority to the 2004 sighting of element 113 by a team of Japanese scientists at the RIKEN Institute in Tokyo (discounting earlier claims of a 2003 discovery at Russia's JINR). 2016 will see a new name suggested, and element 113 will become the first element to be named in Asia.

Ununtrium comes as a pair with ununpentium (see page 398), since it is created when element 115 decays radioactively. The most stable known isotope, Uut-286, has a half-life of 20 seconds, but the heavier the isotope, the more stable ununtrium becomes, and a predicted Uut-287 isotope may have a half-life of as much as 20 minutes.

C Uut

Atomic radius: Unmeasured

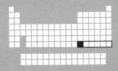

Group 13, Period 7
Transactinide

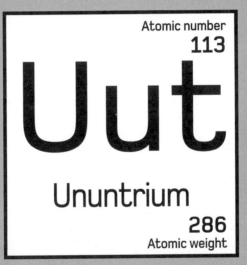

Atomic number
113

# Uut

Ununtrium

286
Atomic weight

Melting point: 427°C (801°F)
Boiling point: 1127°C (2061°F)
Density: 16 g/cm³

Common isotopes: None stable
State (at STP): Unconfirmed
Colour: Unknown

# Flerovium

The 14-year gap between element 114's synthesis in 1998 and its eventual addition to the periodic table in 2012 ranks as the longest wait for confirmation of any element so far. In naming it, the Russian team from JINR honoured Georgy Flyorov (1913–90), head of the Soviet nuclear programme and founder of the institute where element 114 was isolated.

Flerovium is the heaviest element of the carbon group. It was once thought that the element would form the heart of an 'island of stability' (see page 406), hypothetically ranging from element 112 to 118, and have a half-life of millions of years. However, when scientists at JINR fired calcium nuclei into a plutonium target they uncovered an isotope of Fl-289 with a half-life of just 30.4 seconds. Compared to many superheavy elements, this is still the epitome of stability. Predicting flerovium's bulk properties is sport for nuclear physicists: some think it could be a liquid metal, like mercury, while others predict it might be the periodic table's only metallic gas.

C  Fl

Atomic radius: Unmeasured

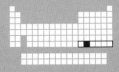

Group 14, Period 7
Transactinide

Atomic number
114

Fl

Flerovium

289
Atomic weight

Melting point: 67°C (153°F)
Boiling point: 147°C (297°F)
Density: 14 g/cm³

Common isotopes: None stable
State (at STP): Unconfirmed
Colour: Unknown

# Ununpentium

The Holy Grail of superheavy element research is to locate so-called islands of stability (see page 406). These are magical 'sweet spots' of the periodic table where the nuclear configurations of protons and neutrons are inherently stable. While all isotopes of such heavy elements would still be radioactive, some might have half-lives numbering in the millions of years, like those of thorium and uranium. However, efforts to find traces of them in nature – or in the detritus of nuclear reactions – have proved unsuccessful, which is why superheavy elements must be labouriously synthesized atom by atom.

In 2004, a Russian-US collaboration (part of the detente between the two sides after the 'transfermium wars' – see page 372) produced four atoms of ununpentium-288. This very 'unmagical' nuclear configuration of this isotope meant it decayed in a 100-millisecond twinkle. Confirmation by IUPAC of the discovery means that in 2016 ununpentium (literally '1-1-5') will receive a new name and be formally installed in the periodic table.

C ⬤ ◯ Uup

Atomic radius: Unmeasured

Group 15, Period 7
Transactinide

Atomic number
115

# Uup

Ununpentium

289
Atomic weight

Melting point: 427°C (801°F)
Boiling point: 1127°C (2061°F)
Density: 13.5 g/cm³

Common isotopes: None stable
State (at STP): Unconfirmed
Colour: Unknown

# Livermorium

On 30 May 2012, two new elements were added to the periodic table: element 114, flerovium, and element 116, livermorium. One unintended consequence of these scientific formalities was to give the periodic table knick-knack industry a shot in the arm, since at one fell swoop everybody's periodic table mugs, ties and posters were rendered out of date. Livermorium is named for the Lawrence Livermore National Laboratory, USA, which collaborated with Russia's JINR to produce several isotopes of element 116 by crashing high-energy calcium nuclei into a target of curium.

The very first atom of livermorium, Lv-296, was synthesized in 2000. The most stable isotope so far created is Lv-293, with a half-life of 61 milliseconds. Understandably, then, this new element's behaviour has not been determined experimentally. It may be the heaviest chalcogen (see page 146), but would it be a heavier version of polonium or would the relativistic effects of such a heavy atom radically alter its chemical character?

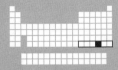

C ◯ ⦿ Lv

Atomic radius: Unmeasured

Group 16, Period 7
Transactinide

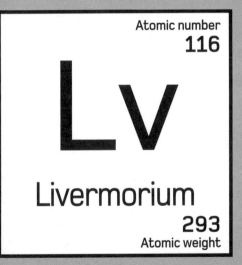

Atomic number
116

Lv

Livermorium

293
Atomic weight

Melting point: 435°C (815°F)
Boiling point: 812°C (1494°F)
Density: 12.9 g/cm³

Common isotopes: None stable
State (at STP): Unconfirmed
Colour: Unknown

# Ununseptium

Superheavy elements are made by nuclear fusion – literally accelerating atomic nuclei and slamming them into other, stationary, nuclei. 'Hot' fusion reactions involve light, high-energy projectiles and heavy actinoid targets. The compound nuclei created have a lot of energy and must evaporate several neutrons to 'cool down'. However, more often than not, they undergo fission reactions and break apart. Thus, the production of new transactinoid elements is a numbers game. In 2010, an international effort between the JINR in Dubna, Russia, and Oak Ridge National Laboratory (ORNL) in Tennessee, USA took 150 days of bombardment to create a single atom of element 117. It had taken ORNL nearly a year and a half to produce a berkelium target weighing 22 mg, but the shipment fell foul of Russian customs. The target criss-crossed the Atlantic Ocean five times, and with a half-life of 330 days, the team at JINR had to hustle to produce Uus-293 and Uus-294 isotopes. Formally recognized in 2015, the discoverers will suggest a new name in 2016.

C  Uus

Atomic radius: Unmeasured

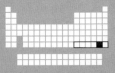

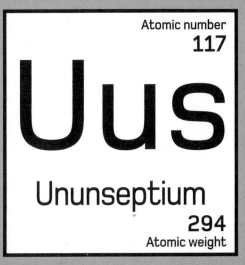

Atomic number
117

# Uus

## Ununseptium

294
Atomic weight

Melting point: 450°C (842°F)
Boiling point: 610°C (1130°F)
Density: 7.2 g/cm³

Common isotopes: None stable
State (at STP): Unconfirmed
Colour: Unknown

# Ununoctium

Creating new superheavy elements is an exercise in chasing the ghosts of atoms. These staggeringly unstable nuclei are produced in single-digit quantities, and often only persist for tiny fractions of a second before self-destruction. Researchers are thus reduced to sifting through a mess of decay-chain daughter isotopes to try to discern the shadowy outline of their elusive quarry. Element 118 is the quintessential will-o'-the-wisp element. It takes something of the order of 10 billion billion collisions of calcium nuclei against a californium target to produce a single atom with an estimated half-life of 0.89 milliseconds. Small wonder that since 2005, only three or four atoms of Uuo-294 have been detected. At the time of writing, the place of element 118 on the periodic table is still occupied by a placeholder element name – ununoctium – and its 'Uuo' chemical symbol. However, priority has been given to a joint Dubna–Berkeley team and, with it, the official naming rights. With the addition of the heaviest of the noble gases, period 7 is complete. But is this the last element of the periodic table?

C ⬤ ⬤ Uuo

Atomic radius: Unmeasured

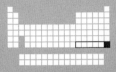

Group 18, Period 7
Transactinide

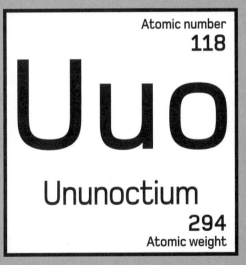

Atomic number
118

# Uuo

Ununoctium

294
Atomic weight

Melting point: −15°C (5°F)
Boiling point: −10°C (14°F)
Density: 5.0 g/cm³

Common isotopes: None stable
State (at STP): Unconfirmed
Colour: Unknown

# Future discoveries

The nuclear shell model, formulated in the late 1960s, proposes that protons and neutrons fill shells in the atomic nucleus in the same way as electrons form shells outside the nucleus. Just as full electron shells give chemical stability, so elements with closed 'nuclear shells' are far more durable against radioactive decay. These so-called 'magic' and 'doubly magic' numbers of protons and neutrons explain why isotopes such as helium-4, oxygen-16, calcium-48 and lead-208 are abundant in the universe.

When elements beyond atomic number 102 proved too fission-prone to synthesize easily, Glenn T. Seaborg suggested there was a 'sea of instability', after which an 'island of stability' stretched from elements 112 to 118. The stability of such nuclei has proven elusive, however, but recent theoretical work predicts an island of stability around unbibium-306 (atomic number 122). The question remains open: is there more to be discovered past element 118? It seems there is still the possibility of some truly unusual chemistry beyond the realms of the current periodic table.

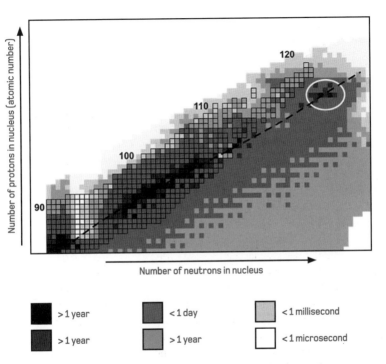

This 'map' of heavy isotopes uses colour coding to identify the measured or calculated half-life of isotopes. The predicted 'island of stability' is ringed.

# Glossary

### Acid
Sour-tasting aqueous solutions that react with bases and metals by donating protons.

### Allotropy
The property of certain chemical elements to have two or more different stable forms, in the same physical state.

### Atom
The basic unit of chemical elements. Atoms are composed of smaller subatomic particles – protons and neutrons in a central nucleus, and electrons in orbitals around the nucleus.

### Atomic number
The number of protons in the atomic nucleus of an element.

### Atomic weight
Also known as relative atomic mass, this is a measure of the average mass of atoms of an element. It works by comparing an element to a theoretical mass of 1/12 of a carbon atom. The average depends on the abundance of isotopes – the differing numbers of neutrons skew the value, so hydrogen with a mass number of 1 has an atomic weight of 1.008.

### Base

Bitter-tasting substances that neutralize acids, reacting with them to produce salts. Generally, bases are substances that accept protons. A base that dissolves in water is called an alkali.

### Compound

A substance or molecule composed of atoms from two or more different elements.

### Earth

An archaic term referring to a stable chemical substance. Earths were once thought to be elements, but are now known to be compounds, often oxides.

### Electronegativity

The tendency of any atom to attract electrons to bond with it. Fluorine is the most electronegative chemical element.

### Electron orbital

Also known as an atomic orbital, the orbitals are the zones within an atom where electrons are most likely to be found. There are multiple orbitals in any given atom and each is associated with a specific energy. In atoms, electrons are not found outside orbitals.

### Element

A substance whose atoms all have the same number of protons in their nuclei, and therefore have the same chemical properties.

### Half-life

The time taken for half a sample of a given radioisotope to decay into a different form through radioactivity.

### Ion

An atom or group of atoms that carries an electric charge thanks to an imbalance between the numbers of negatively charged electrons and positive protons.

### Isotope

An atom of an element with a particular number of neutrons. Isotopes of a given element all have the same atomic number (number of protons) but have different mass numbers (protons + neutrons).

### Lattice

The 3-D organization of atoms within a crystal, made up of repeating units.

### Magic number

The number of nucleons (protons or neutrons) that fills completely a shell within the atomic nucleus. Widely accepted magic numbers are 2, 8, 20, 28, 50, 82 and 126.

### Mass number

The number of nucleons (protons and neutrons) in an atomic nucleus. This is not the same as the measured mass of an atom, or its atomic weight.

### Mineral

A naturally occurring inorganic

solid with a regular crystal arrangement of its atoms.

**Molecule**
An electrically neutral group of two or more atoms linked by chemical bonds.

**Oxidation state**
A number that represents the number of electrons gained, lost or shared by an atom during chemical bonding.

**Polymer**
A large 'macromolecule' made up of repeating subunits. Polymers can be either natural or synthetic molecules.

**Radioisotope**
An isotope whose nucleus is unstable and prone to disintegration through radioactivity.

**Salt**
An ionic compound that forms when an acid reacts with a base, a metal or a molecular ion, or a metal reacts with a non-metal. Stable salts are sometimes called 'earths'.

**Suspension**
A mixture composed of two phases – solid particles suspended in a liquid. Suspensions will eventually settle out under gravity.

**Valence electrons**
The electrons in the outermost orbitals of an atom; those that can participate in bonding.

# Index